Bondage, Gunshots and Mergers

Nine Years on the Road with PeopleSoft

Peter Wortham

authorHOUSE™

1663 LIBERTY DRIVE, SUITE 200
BLOOMINGTON, INDIANA 47403
(800) 839-8640
WWW.AUTHORHOUSE.COM

First published by AuthorHouse 12/22/04

ISBN: 1-4208-2395-7 (sc)

ISBN: 1-4208-2394-9 (dj)

Printed in the United States of America

Bloomington, Indiana

This book is printed on acid-free paper.

Dedication

"To my wife and children for their support
and love over the years. I still don't know why
you put up with me, but I appreciate it to no end"

Thanks also to all the brilliant and funny
consulting friends who became my extended family over the years.
No one can appreciate how bad life can be on the road
like a software consultant.
This book was written for you.

Foreword

This project was actually started two years ago while working in New York and staying at the Marriott Financial Center. I had finished all my email and didn't really have anything pressing in my in-box as it were. To make things worse, there was nothing but crap on cable that night. I remember laying back on a wedge made of pillows and staring at the ceiling with all the lights in the room still on. Thinking back to some of my first consulting assignments and the comedy of errors that seemed to follow me and other consultants everywhere, I thought that an anthology of those events might make for a pretty good book. It started out as a series of short stories about certain assignments but grew into this book you are now holding in your hands. It turned out to be a story about both life on the road and life off the road and some of the challenges surrounding each. It was in the end, written for the traveling consultant in appreciation for their humor, the hardships they endure, the hard work contributed and the headaches suffered.

The content for this book was taken from personal experiences starting in 1995 through 2004 while employed at PeopleSoft in their consulting organization. The character names in the following pages have been changed to protect their identities while customer references have been re-written to reflect only the spirit of the situation, and do not represent the actual people or companies involved. Descriptions in this story will also reflect personal opinion on events as they took place and parts of the story were embellished for the sake of readability. This book is not endorsed by or affiliated with PeopleSoft Incorporated or PeopleSoft Consulting in any way.

Table of Contents

Chapter One - October 1, 2004

Friday, 8AM – Reykjavik, Iceland

While lying back on the narrow twin bed of our tiny but typical hotel room near the city center, my eyes had just started to close when I was startled by the sound of television channels changing in the background. Don, my buddy from Chicago and current traveling partner had zeroed in on the European version of CNN headline news. Still groggy after the long night flight from Boston, I felt myself drift off again until I heard the words, "And PeopleSoft today announced the removal of CEO Craig Conway. The board of directors has named company founder and former CEO Dave Duffield as the new chief executive officer effective immediately." "Wall Street analysts feel this could clear the way for Oracle to complete its takeover."

No caffeine needed, I was fully awake. "In other US news, the latest presidential debate saw a major shift in the post-debate polls showing John Kerry gaining ground, and Mt. Saint Helens erupted again spewing ash and hot gasses thousands of meters into the air." Overall, a slow news day.

The ramifications of the Conway / Duffield swap didn't really start to sink in until I had gotten some sleep, but my mind was racing with all that it might mean. There would be nobody to call and no way to find out what was going on until we returned to the US that following Monday. I laid my head back down on the small flattened pillow and contemplated the news.

There never seemed to be enough time for the mind to evaluate the events that were taking place and then process how to best deal with them. You got to a point in the software consulting business, where survival meant to keep running. New things to learn, new challenges to surmount, new bosses to please, and new customer problems to fix. There wasn't a whole lot of time to sit back, relax or reflect on anything professional or personal.

In this same time period, the never ending hostile take over attempt by Oracle was a constant reminder that our careers could end at any time. It felt like a bad dream that occurred each night, while the Oracle offer deadlines kept passing then endlessly renewed. The employees all knew, or at least suspected that the only way Oracle could realize any margin from the buy-out was to take advantage of the license maintenance revenue and dump the salary expense. For those of you keeping score, that meant most of the PeopleSoft employees. Safra Katz from Oracle had even estimated the PeopleSoft terminations around 6000 when early statements were released to the press.

Other more personal challenges had started to pile up during the same time frame. There was turmoil everywhere. Exhausted in the morning hours at the "Hotel Reykjavik", I started to think back to the last nine years of hotels with bad room service and a constant struggle to remember rental car colors, room numbers, customer vice president names or my own kids birthdays for that matter. The time simply flew by like nobody's business. After 25 states and 7 countries, the artificial glamour of the job usually condensed down into a black and white crime scene photo of a 44 year old male alone in his underwear, on a hotel bed with remote control in hand. Its not an attractive picture.

Some of those memories caused the corners of my mouth to curl upwards, still others would wipe the smile away. In many ways it has been the most challenging and rewarding period in my life both personally and professionally. Now after being totally immersed in the culture and lifestyle and money associated with PeopleSoft, I struggled to find what the next step should be. It was a matter of not being able to see the future or perhaps more accurately stated, not being able to shake off the dust of the last nine years and move on.

Reykjavik, Iceland was a break from everything. Don had called me out of the blue, "Dude, You like Jazz and Blues, right?" he asked. "What about Van Morrison?" he continued. "What are you doing at the end of September?" He snickered a bit. "Stop" I think I said, "What" was the next monosyllabic response. Don responded "Wanna go to Iceland?" I remained silent, naturally dumbfounded.

For some unknown reason the logic to the trip seemed overwhelming, and I felt like I was speaking to a realtor who was just about to make the million dollar club with this next sale. "It was all good", I started thinking. Jazz and Blues are by far my music genre of choice. Live performances in small clubs are by far the best. Iceland sounded like a great place to tour and experience. Van Morrison hardly ever comes back to the US, and besides…the discounted airfare was a huge plus. "How are the schools and what were last year's property taxes?"

It was a brilliant sales pitch and an excellent idea if for no other reason than to get away from a lot of "work" things. I called my wife and mustered every masculine, testosterone-driven personality trait I could. It needed to be clear who wore the pants in the family. The law was about to be laid down. "Honey, can I go to Iceland with Don?". And yes, it sounded just as wimpy as it reads here.

Of course I am playing up the drama, but my wife and I have a pretty cool agreement about time off and mini vacations. Aside from our family vacations, when she needs a break from the kids (or me) she heads north to the cabin or even off to Chicago or New York with her sisters for a shopping weekend. When it comes to a golf weekend for me or a chance to do something with friends, its never a problem. From what I can see, more couples need to adopt this fair-trade policy. "Thank you dear". Tickets were booked and Don and I made plans.

Don and I both had similar experiences with the consulting world and it was good to talk with someone who could listen to me vent just as easily as I could listen to them vent. Don had been at PeopleSoft years before but left to pursue other opportunities a couple of years ago. We both had reservations about the nonstop travel especially when trying to maintain a relationship or family. We traded stories of nightmares and we laughed about a common experience in Manhattan where we first met. It was good to blow off some steam, and see Van Morrison in a pretty cool tour-stop at the same time.

It was a revealing discussion and we both struggled to pinpoint exactly why we still did this for a living. How did we balance the whole work/life thing? What things in our lives were truly important to us? Were metal woods really still woods if they were made out of steel?

That in a nutshell was how our minds worked. Serious questions followed by declarations of utter nonsense.

The weekend was extremely therapeutic, offering Don and I the chance to share a few stories and of course a few local beers too. The challenge of maintaining a personal life with our current jobs was a common topic throughout the weekend. My family and I were close or so I felt it was true but I remembered that my view on what was considered "normal" in terms of relationships was probably not normal at all.

We enjoyed the whole weekend, the sights, the clubs, the food and the people there. Van Morrison put on a terrific show where much of his tone and delivery in the concert was very R&B, befitting my mood. The crowd in the small auditorium seemed very reserved, applauding politely at the end of each song but I wondered what Van thought of the crowd response. He admitted on stage that it was his first time performing in Iceland, but these people seemed very conservative during the concert. Ah well. It was an excellent event.

Don and I packed up our things Sunday morning, including some artwork we bought on the streets the day before and made our way back to Keflavik and the international airport there. Customs was reasonable and easy enough to get through while the airplane was actually sitting at the gate. Both were good signs. We were still a bit groggy from the concert and the bar from the night before, and it was very likely that we both would fade as soon as the wheels were up.

I started to miss my wife and looked forward to getting home again to see her and the kids too. As the plane lifted and the G's pulled my head backward into the headrest, I thought back to how I got started with this company and why I loved the work so much. How much travel had I endured and how much time did I actually spend away from the family?. How did I get to be 44 when I was just 30-something a few minutes ago, and how did I manage to gain 60 pounds in the process? I could feel my eyebrows furrow downward at first but eventually easing back as everything faded softly to black.

Chapter Two - 1995

Stan Chamberlain positioned himself towards the stage at the front of the auditorium so he could revel in the looks of anguish flowing over the faces of the people under his command. They had been called together like this only twice before in the past 5 years and then only to receive news of department wide layoffs or cuts in benefits. The tension in the room was as thick as Stan's moustache as he used the palm of his hand to "comb" it down before he spoke. Stan was one of the two managers in the technology department to wear a suit to the office as a way to punctuate his authority. Everyone else in the department was required to wear ties but coats were reserved for executives. Stan preferred black pants and a gray hounds tooth coat above a standard two piece.

Climbing three steps to the top of the stage, Stan walked slowly to the center. Without any expression whatsoever, Stan swept the crowd with his eyes to indicate he was ready to deliver the bad news. The room fell immediately silent, and Stan could not hide the little grin that emerged at the control he possessed over the mass of emotion in the room.

"As you may have heard, we have been in negotiations with our parent company over the potential sale of our life insurance division". I heard the sound of a lung-filling inhale behind me and a sigh to my left. "Our parent company has chosen to accept an offer for our division from a larger and healthier company based out of Charleston, South Carolina." "It will be an equal merger." That was all that really needed to be said, because at that point my mind simply took over to run through the likely scenarios for the next few months. Stan kept talking and people in the room sat there silently, in total shock.

The victim of small regional bank consolidation, I had experienced this feeling 5 years earlier. The bank I used to work for was an easy buy out target owning a steady commercial customer list, low overall dept and a large credit card customer base. The stockholders chose to take the deal when a larger competitor made a generous stock offer. That was about it for all of our jobs in the technical support area.

The words coming out of Stan's mouth now, were all words designed to reassure the masses in the room, but I had heard them all before, right before the staff was systematically reduced bit by bit. And it was done a little at a time for a reason. News of small reductions in work force never make it to the papers, and it also tends to influence others to leave on their own. It has become a nice little repeatable process that you pick up on once you've been through it. The writing was on the wall.

Stan was as eloquent as a working class kid from Detroit could be, while he dispensed the misinformation he was instructed to sell. This was an 'insurance' company after all and this was another sales pitch. There was talk about expanding the data center in Michigan so that our combined companies could expand together. Stan went on about how impressed the Charleston company was with the experience and expertise of our programming staff. Yada, Yada, Yada. Stan went on for 90 minutes.

I would have expected the next morning after a meeting like Stan's to be morbid around the office but not today. People in the "bullpen" as the programming floor was called, were talking about the positive effects of "merging" with a larger company. "Like Stan said, they were looking for a Northern company with a data center as well." "We're a perfect fit for them." Stan's plan was coming together. The folks here actually believed the unadulterated bullshit that came out of his mouth the day before.

Johnny Badanczyk and I were too busy working on our own software project trying to make our go-live date to bother with anyone. We had been pumping in the hours willingly to make the implementation a success, mostly because the software was "fucking cool" in John's words. The PeopleSoft GL Release 2 package we were customizing was completely different from anything else used in the company.

Our insurance company was purely a mainframe based, COBOL processing shop. All the transaction processing was done at night, in huge batch jobs and supporting it was about as interesting as scraping dirt from under your fingernails. The "PeopleSoft" project was based on completely different technology at the time, new to the entire marketplace. It was an on-line graphical system that ran on a Windows desktop with an Oracle database as its heart. John was right. It was "fucking cool".

Well into the project timeline, we had hired a consultant from PeopleSoft to help us out with a couple of configuration issues between the database server and the desktop. He was only available for a few days before having to jet off to another customer site. While there we talked about how the PeopleSoft company was growing and how he couldn't keep up with the never ending requests for technical consulting. He also talked about the money. Indirectly of course, nobody really ever talks about what they make or how big their house is but it was his little comments that stuck with us. Last minute trips off to Europe with his wife, wanting to find a slightly better deal on the red 911 he was looking to buy, needing to get back to New York in the fall to find good deals on men's dress shoes.

"Maybe this guy was dropping hints for a reason" John had commented after the consultant had left. When you're struggling to make payments on a tiny house and saving a dollar here and there to buy clothes for three kids, your ears tend to perk up when you hear stuff like this. Suddenly being a mainframe COBOL programmer for an insurance company about to be "downsized" wasn't all that appealing.

Independently, John and I made a few calls to the folks we knew at PeopleSoft on the customer service side of the company. They were helpful and interested but they were quick to point out that "contractual obligations would prevent PeopleSoft from hiring anyone from an active customer of theirs." "We're not allowed to approach you." In other words, as long as we worked for a contracted customer, we couldn't work for PeopleSoft.

We would chat with our customer service guy over the course of the next two months but nothing really ever came out of it until the next major announcement by Stan the "man". A company memo was issued to announce the legal merger date where we would all become official employees of the "Great Carolina Life Insurance Company". "A day we look forward to with pride" the memo said. More importantly, a day under which all software and hardware support contracts terminate with the old insurance company. John and I however, did not yet understand the significance of this date.

A couple of days later while still trudging along with testing the software changes we worked hard and long to make, I got a call from our friendly PeopleSoft customer service guy, Kyle. "How are you guys doing" he asked. "Pretty well." "We're about through with testing and aside from a couple of performance problems with the database we're feeling pretty good about putting this thing into production." "That's great" Kyle answered, "I want to make sure you guys get the product into production, then we can talk." "Talk about what?" (I was so naïve back then), "Well" Kyle took a breath, "When your company completes the merger, that's it for our contract with you." "It all needs to be re-negotiated with the governing merger partner which in this case is not you". "We're free to have discussions about employment at that time and we're very interested, but you have to make sure nobody leaves until the project goes live."

There was a pause on the phone and I didn't rush to say anything until Kyle jumped back in….."We want to show that you can finish a project (cradle to grave) so you don't want to put the project in jeopardy when you leave." "We also want to avoid litigation if at all possible".

It was all too much for me to absorb at first. Somebody from my old employer suing another company because of me leaving or because of some commodity I now possessed? I couldn't fathom it. "Listen, thanks for all the information and I very much appreciate what you just told me. Talk to you soon." I hung up the phone and immediately began to swell with a feeling of "importance" for lack of a better word. If it were up to Stan, your 18 hour days and weekends spent in the office to learn new technologies were "just part of doing your job". There was never a pat on the back, much less monetary reward for "killing yourself" on a project. It was hard to believe that somebody actually wanted me to leave and join their company. I was being recruited. It was a "first" for me. I felt so "special", but not in a short bus kind of way.

At home I broached the topic carefully with my wife because one of the underlying requirements for the job Kyle and I were "not talking about" was that it required travel. I wasn't sure how much, but I would be doing the kind of on-site, out of state customer support as a big part of the job. It was a tough subject. Our youngest daughter was about two years old and the other two kids were more than a handful. How would things be if I weren't around to help? We talked about the money too. We were a family who could pay their bills but who always needed the next paycheck for something. I hated having to plan a month or more in advance to replace two of my 4 bald tires. I remember having to call home once to see if I could hit the ATM for $20 so I could go with the guys for a beer. Oh, and forget about saving any money for retirement or college or vacation. It sucked.

There was really only one room in the house. The "front room" couldn't really be called the living room or the great room because it wasn't big enough for living, nor was it great. It was typically littered with toys and impossible to walk through. We had a small kitchen and an eating area that met the front room through a double door opening. Down the narrow hallway were the three bedrooms and a single bath.

When the three kids were confined to the house in the cold winter months, the concept of a "play area" meant the whole house could be destroyed in a matter of minutes by three munchkins all under 4 foot tall. We lived in 900 square feet of pure bliss. "Hey, stop washing Barbie's hair and get out of the bathroom, I have to pee."

We needed more room, we wanted more money, banks and insurance companies and COBOL programming were not going to get us there.

Kyle knew exactly where I was coming from. Although he came from a relatively well-to-do background, he had heard a similar stories from other poor bastards that were in my situation. It was all too easy for Kyle to pluck guys like me from our meager surroundings and offer them a chance for a better financial life. A better "quality of life" was a different story and the central meaning of the term was up for interpretation.

By early September, John and I had all the kinks worked out of the GL system and it was running in system-test mode smoothly. The functional project team was running reports and on-line traffic in parallel to the old system to validate that the new system would be in balance. It was a beautiful thing to know we had beaten this thing down with virtually no knowledge about the product, the database or the hardware architecture. Alan from the DBA group had been awesome on the database management and tuning effort and Don had been a stand up guy with all his help on network and file system management. We were a tiny but effective technical team.

One night while babysitting a few data upload processes, John and I decided to bring in our roller blades and skate in the parking lot after hours. Everyone else had gone home for the day and Stan may have even been the first one out the door. Don was working late but on his way out when he saw us trying not to fall down in the back parking area. He offered to stop for a little liquid refreshment and returned 15 minutes later with some cold "Zima". Zima was very cool at the time. (What were we thinking).

We sat on the grass by the curb and drank while the parking lot lights came on. We talked about frustration and opportunity and everyone was candid with their thoughts. We were all ready to make a move before the walls came tumbling down.

The following week our new "combined" management team pulled the plug on our PeopleSoft project. Everyone on the team was crushed especially since we were so close to go-live and with everything in balance to the system we were replacing. It is hard to measure the level of disappointment after 6 months of long hard effort. "It is what it is and nothing more" spouted someone from the functional team as a group of them had gathered near us to talk about what they had heard. It was all wasted effort we thought, but the project cancellation made it a little easier to make a tough decision on a different front. My interview with PeopleSoft was scheduled for the following week.

Mike Maszka was the regional director for the consulting group at PeopleSoft at the time. I made the trip to Chicago to meet with him only to wait in the reception area well past my scheduled interview time. I felt a little like a goof sitting there in my new suit only to see everyone walk through the lobby in faded jeans and t-shirts. One guy was wearing shorts and Birkenstocks. "What is this place?"

Mike came out with a smile and finally directed me to his office down the hall. It was a pleasant conversation but he seemed to be concerned about my lack of functional product knowledge. I told Mike that I/we focused on the technical aspects of the project and that there was much to absorb just in those areas. We had performed a bunch of Oracle database tuning on table spaces and indexes, as well as managing all the code changes.

During our testing, I found out that our installation represented the largest GL transaction processing customer for PeopleSoft. We had broken new ground with the volumes of transactions we were pumping through the system, but Mike already knew that. He was interested in me I think, because newer customers about to sign software contracts were going to push the envelope even further. Of course I knew none of that then, which allowed Mike to hide how interested he might have been.

The interview lasted only 20 minutes which is normally a bad sign. We had a pleasant conversation but I figured that it would be longer if for no other reason than I drove 6 hours to be there. My final impression was that it must not have gone well even though I knew they needed more technical people with experience on the team. The drive home was incredibly long and void of a single positive thought.

A week later I received a call from the HR department at PeopleSoft wanting to verify my address for an offer letter. "Uh, offer letter?" I responded like a teenager on "Lewds". "Yes, didn't Mike call you to extend you a verbal offer?" the pleasant female voice asked. "No he didn't, but can I ask what the content of the letter says?" She went on to discuss some of the details where the highpoint was a pure doubling of my current salary plus a quarterly bonus plan that put most insurance company director salaries to shame. "I would be making more that Stan." "How cool is that?".

I had figured that Mike must have had a lengthy conversation with Kyle about the work I had done at the insurance company, and that the interview was more of a formality than anything else. I was right, Kyle confirmed later that internal referrals are the primary way PeopleSoft found its new employees. "We need to know as soon as possible" the HR lady said, "Our next boot camp is scheduled in two weeks and you would be slotted for a position in that class". I confirmed my address information and promised to call back with an answer in the next couple of days. "Boot Camp?", "What Boot Camp?"

Stan actually took the news pretty well, but he felt compelled to ask if the reason for the job change was for "money or opportunity". "Both really" I replied, telling Stan about the new technology I would be getting into on a regular basis. I was also hoping that he would ask about the money, and finally he did at the end of our discussion. I told him. "Well it's a pretty safe bet that you're not going to get a counter offer from us on that number." It was pure joy. As I walked back to my desk I remember mumbling a phrase I had recently learned from a friend who had lived in Buenos Aires;

"Come' Mierda y' Ladre a la Luna". Eat shit and bark at the moon, Stan.

Cheryl and I immediately started looking for a newer house with more room for the kids. "TWO BATHROOMS" she added.

Chapter Three – Bootcamp

You just have to love a moniker for corporate training called "boot camp". There seems to be a suggestion that some of the "recruits" may not make it out alive. It gave me something to think about and maybe worry about on the long plane ride from Detroit to San Francisco where the company headquarters were located. We were all given maps and instructions on where to go, where to stay and what to wear. The whole of the experience was a little scary for me, never really being away from home alone. I've gone on business trips with other guys from my former company but this was different, exciting and a bit intimidating at the same time.

This mid October morning in San Francisco was very pleasant compared to the 45 degrees and drizzle I left back in Michigan. I guess I didn't need my down jacket after all. Following delivered instructions, I rented a car to make my trek to the PeopleSoft offices just outside of Pleasanton. The Hertz agent was friendly and offered advice on how to find the freeways, but I felt the urge to say something like "I'm from Detroit, home of the car, I don't need directions". Of course that would have sounded completely foolish and I was trying hard not to look like a complete idiot from out of town. I said "thank you", grabbed my attractive 10 year old powder blue vinyl luggage and headed for the Hertz courtesy bus.

15 minutes later I found my midnight blue Ford Taurus, opened my highlighted Hertz map and proceeded to highway 101, South. The Campus wasn't that hard to find and it felt good to find my way around a strange town without my wife there to do the navigating. She loved to navigate, but don't most women?

I parked in a nice sunny wide-open spot in the middle of the fresh black asphalt parking lot which was an automatic reaction for typical October days in Michigan. I would learn later that this was probably not a good idea in California for a dark sedan with a dark interior and a black vinyl steering wheel. I happily grabbed my orientation folder, rolled up my windows extra tight, and headed for the training center on the 5th floor.

Everyone in the reception area was pleasant and I found the elevators to the right, gleaming with fingerprint-free chrome. My kids would have loved this fresh canvas for dirty hands. I missed them already. We all met in a conference room with a large oval table that sat 16 and I walked through the door to be greeted by three people simultaneously. "Hi, how you doin'" was the best I could apparently muster under the moderate stress I was suffering at the time.

In retrospect I should have been ready with a greeting that sounded a little more intelligent. Conversations were started easily in this group and all seemed to be extremely happy to be there. A few others came in 5 minutes later followed by a young lady carrying a box load of small binders.

Cindy Randall was 5 foot 5 with blond frizzy hair pulled back from her face with a black headband. She like most of the people we had met from PeopleSoft so far, was wearing faded blue jeans and a simple short sleeve light weight sweater. It was completely casual here and it helped to put everyone in the room at ease instantly. Cindy, welcomed us with a huge smile and passed out the binders. She talked a hundred miles an hour which was fine, because everyone seemed to be feeding off this heightened sense of awareness. Maybe it was the sugar and caffeine from coffee and doughnuts in the corner of the room.

We received applications for corporate credit cards, international calling cards, and gold services at rental car companies. Cindy told us our laptop computers would be delivered to us before the end of the first week of class, and the binder included all sorts of information for the HR department, payroll, technical support, you name it. It was the most attention and the biggest investment of time and money any employer ever made paid to me.

I felt that all the thankless work I performed over the years had finally paid off. Somebody recognized that I, and everyone else in the room had some potential. Cindy put together a couple of presentations for us about stock plans and 401K benefits and all of the things I was never offered anywhere else. My head hurt after 3 hours of endless benefits offerings. Cindy let us go for a lunch break and pointed us in the direction of a large grass courtyard and a well stocked cafeteria.

I normally have to be careful when I meet a new crowd of people because as much as I want to be an "average" normal guy who just wants to fit in, I'm actually 6 foot 6 and stick out like a sore thumb. I've noticed that when I get excited about anything and start to use my arms to emphasize words (I have a colorful European ancestry), people tend to back away. Everyone in my family talks with their hands, it isn't pretty.

It seemed that many of the people in our class had the same familial background. The arms were flailing everywhere as we walked to the cafeteria. Most of the men were pretty good sized as well. All over 6 foot and in pretty good shape, I wondered if there was some sort of minimum height requirement for working here. I didn't necessarily feel out of place, until the conversations picked up in detail. Kim , Randy and Karen had all gone to Stanford but separated by a few years, Tom and Vincent had attended Brown in the east. There were a couple of people from Purdue, two from the University of Michigan, and few more from Columbia.

I kept my mouth shut about my little technical college in the Detroit suburbs. The leveler though was that everyone has risen up from the ranks at whatever company and from whatever industry they came from. Everyone seemed to have been responsible for some major role in a software implementation project and also had experience with PeopleSoft products. I could speak about our project and the technical role we all played there, and that we pushed the edge of the envelope with the volume of transactions we cranked through the system. After the technical guys heard how we did it, I was immediately accepted into the group.

Later that day I got a call from Johnny back at the Insurance company who had been making contacts with PeopleSoft for the last two weeks. It looked like I would miss John for this particular boot camp, but he would be happily following soon. Though I didn't know it at the time, Don was looking into making the jump and John and I also heard about one of our other project buddies Gail, who was asking about the company too. We would ultimately move 5 from our project team over to the "dark side" at PeopleSoft as John liked to call it.

Walking around the PeopleSoft campus in 1995, I noticed that pretty much everybody was in jeans or shorts. The 3rd floor of our ATT leased building was mostly reserved for the sales group. Business casual was the mode on the third floor. I remember looking for someone from PeopleSoft I had met at our insurance company earlier in the year who also resided on the third floor and I walked by rows of identical offices each with the most basic of desks and chairs. I passed by Al Duffield, Dave Duffield, and other names I recognized from our first day of orientation. There were smiles, and laughter, and blue jeans and positive energy everywhere. Quite a change from whence I came.

Eventually I found the guy I was looking for, and his office was simply filled with vendor toys. Yo-yo's, and stress balls, games and puzzles. In one corner sat a chair that doubled as a "twin" sized futon. He noticed me looking at it and said "yeah, I sometimes work here all night". He welcomed me to the company and I shook his hand and we promised to meet for lunch at the cafeteria sometime during the second week of boot camp.

As I walked back out towards the lobby, I stopped at the rest room and took care of all the free soda and coffee that had built up, and Dave Duffield came in to stand next to me at the only open space. I wondered how stupid it would look to start an introduction and a conversation over the mini wall between the urinals. I mean, what do you do in a situation like that, offer to shake the man's hand? I think not. I remembered one of the company "core" values (clean bathrooms) on the way out however, and wiped down the sink area before I left.

The PeopleTools development group was reputed to be housed on the fourth floor and our new electronic cards allowed us to roam all over the building so I roamed. Around one corner on the 4th was a large open room, carpeted and filled with comfortable chairs and one pool table right in the middle. There was a mini-kitchen towards the back and a few magazines and papers there for reading. There were more beards and pony-tails on this floor than I had ever seen in one place. Suddenly my short banker's haircut started to give me away for what I was… either visitor or sales guy. I enjoyed my little tour and headed back to our conference room on the first floor for more training.

Training was tougher than I expected but then again it assumed that you knew something about the product before the detailed training started. At the midpoint of the third week I received a message from Mike back in Chicago that they needed to pull me out of class a couple of days early to help fix some technical issues they were having at a customer in Hartford. I was to fly home (first time in three weeks) grab some business formal clothes and be in Hartford by Thursday morning. I would receive contact names and other logistics information in an encrypted e-mail I could get by dialing in with my laptop. It had everything but "code names" and the Walther PPK. "Bond, James Bond" thank you very much, it was simply a natural high. I said my good byes and headed back to Michigan.

Chapter Four - Hartford

Family reunions take on a whole different meaning when it is you that has been away from your own immediate family. Although it felt like a year or more had passed since I had seen my family, it was still two weeks removed from my kids and my wife and that was enough. It was an incredibly long flight and the anticipation of seeing my family again had me nervously tapping my fingers on my armrest and bouncing my foot on its ball for five straight hours. Cheryl hugged me a little tighter than normal and the kids were acting up just to get a little attention from Dad.

It was good to be home but hard knowing that I had to tell them the bad news of my immediate departure to Hartford in the morning. Cheryl was ultimately happy for me. I couldn't hide the excitement I was feeling around the new job and it was clear that this company was making an investment in its employees where so many previous employers had not. She read over the medical and benefits package that I had carelessly opened and briefly skimmed, but she was pleased to find very good coverage for the family as well as some investment opportunities that were subsidized by PeopleSoft.

She was happy that I was happy and there would be no mention of the kids acting up during the three weeks I was gone or the fact that she worked full time but also had to cut the grass, get the oil changed in the van or struggle through the kids homework alone.

After 10+ years of oppressive employers, she was letting me run with this career change and wouldn't do anything to spoil the upward curves that now frequently appeared at the corners of my mouth.

I had received an e-mail from Janice Markum who was a well respected Account Manager from the Chicago office. Janice went on in some detail about a customer in Hartford who could not make a workstation connection to the database. In the few instances where they could, they complained about poor system performance. It would be my job to evaluate the hardware and technical infrastructure situation there and simply "make things work" as it were. She left instructions about where to meet, in the lobby of the hotel that adjoined a shopping mall and the now "former" Hartford Whalers hockey arena. I booked my last minute reservations with our corporate travel agent "Francesca" and leaving sad faces behind in Detroit, I made my way to Detroit Metro Airport and ultimately Hartford.

The next morning I washed the sleep from my eyes, face, toes and crotch while doing my best to wake up in the hottest shower I could stand. After the whirlwind of the last few days and the general lack of sleep, it was all I could do to assemble my wits about me, get dressed and make it down to the lobby by 7:30 EST. Without much of a personal description from Janice, I told her that I would be the 6 foot 6 guy in the blue suit and the blue computer backpack with the PeopleSoft logo. Good enough. She found me in the lobby next to the red patterned Victorian chairs, looking like a stranger in a strange town.

Janice quickly went into details of the problem at hand. We were apparently here in pre-sales mode. The customer was comparing three software products supposedly side by side to see which could handle the expected general ledger transaction volume through the edit and post nightly processing. Our software was apparently "not working". We discussed who some of the key players were as well as the fact that one of the 'big 5" audit/consulting firms was running the independent software comparison. "Independent" was the word out of Janice's mouth spoken with a twinge of disdain.

She went on to explain that the consulting firm doing the software comparison was also the largest "field" implementer of one of our biggest software competitors. Lets just call them "Williams Software", was one of our challengers at this insurance company and it was apparently running "just fine" under the watchful and impartial eye of the audit firm. "How convenient" we both agreed. It would appear that the audit company was trying to steer the purchase decision towards Williams Software, of course that was only my opinion.

We arrived on-site at 9:00 AM and were shown to a large conference room with 4 long tables assembled into a large square in the center. The perimeter of the room was buffeted by narrow tables with Windows 95 desktop computers every three feet. The entire room was filled with consultants all working on a piece of the software comparison.

Some were writing up summaries of product features, while others were working on running "batch" processes to see how much transaction volume they could pump through the system. I had assumed that because this was all discussed as a "side by side" comparison that I would be able to find the "technical" group who was struggling with our software. There were about 30 people in this room after all, "where are the PeopleSoft consultants?" I asked.

One of the more formally dressed ladies in the room quickly approached me to say hello and to explain that she was "in charge of the team on the ground here".

"Great, nice to meet you".

"I was wondering where the technical team is, working with PeopleSoft I mean." I asked trying to sound semi-professional in the process.

She explained that after numerous problems trying to configure our software and make it work, they had given up. I asked about the functionality of the system after our installer had finished.

"I understood that the customer had signed off on a functionality checklist after our installer was finished here."

"The system would have to have been operational before the PeopleSoft installer would ever leave the customer site." I followed.

The large woman with the tan suit and paisley scarf replied "Well it must have never really worked then because we were unable to connect to the system from any of our workstations here."

Janice stepped in to soften the tone of the conversation, after all that was part of her job too. We left it open to investigation not placing blame on anyone, just focusing on finding out what the problems were and making whatever corrections we needed. I remembered the phrase my daughter would use each time one of her toys stopped working. "Daddy fix?". Yes Janice, daddy fix.

At this point the paisley scarf pointed me to the corner of the room where a desktop computer sat crushed into the corner flat up against the wall with the keyboard laying on top of the monitor. It was off, of course. Out of the 30 people in the room running this "unbiased product comparison", none of them were actively working with our software.

The machine in the corner was moved out from the wall and the box was powered up. I did a little poking around to find that some of the delivered configuration settings and files needed to run our software were missing. "Interesting" I thought, "sabotage?" was my next thought but it was probably more like "reckless endangerment".

I found the lone customer technical guy on site who was really the only contact between the throng of consultants in the conference room and his company's technical people. Twenty minutes of questions with the lone technical guy, we'll just call him "Bob", led me to believe that in his infinite wisdom he may have inadvertently screwed up the whole thing.

You just have to love computer technical guys. They seem to be much like the car guys I grew up with. No matter the true nature of the problem or the inherent risk of taking such an action, there seems to be an involuntary attraction to taking something apart in order to fix it. "I don't know what's wrong Cooter, but hand me that torque wrench and hammer and I'll fix it…"

"Bob" started digging into the files and configuration variables on the workstation and without realizing what he'd done, managed to disable the whole thing in about 30 minutes. This was all done after our installer had left the site and where everything was actually working fine. Poor "Bob" just had to play with that darn thing and bust it. "Damn it Cooter, why did ya let Bobby touch the thing in the first place".

Into mid-afternoon, I managed to get the workstation re-configured with the right files and right paths and we were able to connect to the database after their DBA brought the thing up and active. From there we had a good working model from which we could "stage" or set up other workstations for access to PeopleSoft GL software. It was a matter of cloning the workstation to give the others in the room access.

We were there late setting up other workstations and wrapped up work around 9 PM. Janice was pleased to see that we had figured out one of the major issues and showed the potential customer that we could help them with this kind of technical support. It felt good for me personally because I seemed to pass my first test in the field.

"Interesting town, Hartford". A beautiful new and clean city with a number of high rise buildings scattered to and fro. It's just a little difficult finding anything to do there after 6 PM on a week night. The entire downtown area seemed to shut down.

The streets are empty, and restaurants and shops are closed and it finally occurs to you that there simply is no residential base in the downtown area. No residential base, ya know, people. We found that the hotel was our only recourse for food, and upon arrival somewhere around 9:50 PM, heard that the kitchen would be closing at 10:00. It was a mad dash to find something on the menu and place our order.

No one seemed to be happy to see us walk through the door although the waitress was relatively pleasant. The cook in the back of the restaurant was less than pleased. We could hear the argument in the back save for the specific words. Janice thought she heard something about having to "turn everything back on". I assumed that the cook meant the "grill" which he probably turned off early in anticipation of getting out of Dodge. He cooked, we ate, Janice paid and we were out of there by 10:45.

Day two was a whirlwind. The consulting firm seemed to want to find issues with our software rather than work to perform the side by side comparison. There was no attempt to assign us any of the 30 resources in the room so they all kept trudging along with trying to get their edit and post processes to perform to the standard set by the customer.

I spent some more time with "Bob" to gain access to certain database and server functions I needed to tune our own processes. "Bob" was reluctant but finally agreed knowing that he couldn't really help us and also understanding that we were a bit understaffed in this "unbiased product comparison" at 30 resources to 2.

Working with another DBA that Janice had flown in from St. Louis, we were able to clean up some of the default settings delivered with the software in favor of another configuration more efficient for large transaction volumes. After a day's worth of work and a couple of trial runs, we were able to pump 200,000 transactions through edit and post processes in less that an hour on shitty hardware. In Detroit terms, we made the most out of the situation and got an old tired engine to kick a little ass. In less than 48 hours we had beaten the best processing times of our competitor after they had been working on tuning for weeks.

Janice was all smiles and had wonderful things to say about our little victory. It was clearly a monkey off her back and a great chance for me to get my feet wet in the field. The stress of the past couple days was intense though and I hoped that this was not the normal course of doing business here. I could see my brain caving in pretty easily after a few weeks of non-stop pressure, no food and limited sleep. In any case it was also time to head back home just as soon as I checked for messages. There it was. One message. Another town and another tuning assignment

Chapter Five – Beer and Religion

College towns. You just have to love small college towns. Lacrosse reminded me a ton of Mt. Pleasant Michigan and Central Michigan University. Interestingly enough Lacrosse Wisconsin had more hills than Mt. Pleasant could ever hope for. It was an oxymoron really to give a city a name like Mt. Pleasant when the highest elevation in the whole town is the salt pile they used for clearing streets in the winter. Lacrosse was quite beautiful really with plenty of rolling green hills and older turn-of-the-century downtown area with a brewery smack dab in the middle of the city for good measure.

Mike had called me to let me know that my "tuning" efforts were about to be tested again but this time on another product, Accounts Receivable on the current release 3.0. I was to report to one of the city's largest commercial equipment manufacturers, but one of PeopleSoft's smallest customers. It was apparently a big deal for them to agree to have me come out mostly because of our billing rates.

My rate was $215 an hour and high when compared to other consultants in the software industry, but these folks needed tuning help. It was impacting their ability to do system testing and also preventing them from going "live" with the product.

I arrived Monday and quickly met all the interested parties including the company's DBA and two fine gentlemen from a local consulting company charged with on-going support of our software for this customer. Everyone here was extremely pleasant and happy to finally get someone from PeopleSoft to help them out.

The pressure wasn't as intense as my first assignment mostly because we weren't about to lose a multi-million dollar software sale. Still these folks were having trouble and we wanted to make sure we could get them to production on time.

The first few days were long but they went well. Working with the DBA, I was able to try a couple of things that I had used at the insurance company that would be considered "general" performance tuning. They seemed to improve performance about 20% but wouldn't be enough of a gain to satisfy production requirements. The work went on for a few weeks with steady gains in batch transaction performance every few days.

One night Ralph Engles, the significantly older of the two consulting guys already on-site at the customer, took me out for dinner, drinks and a visit to the historic home he was staying at in Lacrosse. Ralph seemed to dislike his own name because after referring to him that way he stopped me with an outstretched palm to say "Just call me RJ".

I'm sure RJ was short for something that included his middle name but Ralph offered no more explanation and I wasn't really in any position to push the topic further. RJ was about 5' 8"and almost completely bald with light brown hair raked across the top of his head. He was a bit overweight but carried himself well even in a rushed walk down the hallways when late for a meeting. RJ had a round and kind face which was blocked by a pair of dark plastic framed glasses. He didn't wear either, but RJ would have looked perfect in a white short sleeve shirt and pocket protector. Instead, RJ was a khaki and plaid shirt kind-of-guy.

We went to a local restaurant on the outskirts of the city. It had a name that I can't remember to this day but it had to be something like "Mable's Diner" or "Laverne's Luncheonette". The chrome and Formica tables were right out of a 50's sit com complete with red checkerboard paper placemats. It was all Midwestern food here with plenty of butter or bacon grease coating everything served.

I believe I had the pot roast and mashed potatoes which of course were wonderful on a slightly chilly night. RJ made it a point to wait for my reaction to the first couple of bites. "See, what did I tell ya." He was right, it really was good. "Nothin' like comfort food on a cold November night".

From there we drove around a bit and eventually met the other part of the duo from his company, Tim. Tim had suggested that we meet him at a pub closer to the UW-Lacrosse campus. He was still in his twenties, which explained his choice for this evening's drinking establishment. Tim had worked a bit later to finish up on a few things and brought us good news on some other performance gains from changes we had made earlier in the day. It really wasn't a reason to celebrate but we seemed to get along pretty well, so "Why the hell not".

The G. Heilman Brewing Company was all of 2 miles away, which made the beer brand of choice here to be anything made locally at the brewery. "Old Style" and "Special Export" were the two big brands and always on sale here at the "Wayside Tavern". Tim had already ordered a burger and fries and proceeded to turn his head slowly from side to side occasionally to check out the college-babe action until his pub burger arrived.

To fit in with the crowd, Tim wore faded jeans and a dark brown long sleeve flannel shirt topped off with a baseball cap and a downward curled brim. It was very much a Seattle-grunge-thing and everyone in the place was going for the same look. "Annie Hall all over again…. With baseball caps" I blurted. Tim had clued me into the motif of the bar and I came prepared with a Hard Rock ball cap of my own although I'm pretty sure I still looked my age.

We were very comfortable at the bar after a couple of "Special Ex" drafts and we started to talk about the early days of muscle cars and hanging out at the Burger King when we were teenagers, but that was just RJ and me. Tim was looking at us kind of funny as we talked about "Hemi" engines and Pontiac 455 big blocks.

"What are you talking about?"

Tim had made a face downward-turned eyebrows,

"Engines man". "Big fucking engines."

"The kind that would scramble an egg in a glass sitting on the dashboard while the motor idled." RJ was waxing poetic.

Tim tried to participate with "I had a Celica with a sweet V-6". RJ and I just laughed.

After a half hour or so, the bartender moved to the end of the bar to reach for a small length of yellow rope attached to a brass ship's bell. He yanked on the rope three quick times then shouted out to the crowd "TEN CENT BEER FOR THE NEXT TEN MINUTES!". I expected a mad rush up to the rectangular bar in the center of the room but only a few people came up to get their little 10 ounce shell of "Old Style".

RJ and I looked at each other and both shook our heads in unison. We were both thinking that the 50 or so people in the place would be rushing up for ten-cent beer, I mean these were college kids after all. Where's the college pride? RJ figured that because it was a weeknight they must have been taking it easy.

I called to the bartender and asked him the same question but got a different answer. "It's usually a money thing for most of these guys, breaking a buck is a hard thing to do when its lunch money for tomorrow…". I then told him "Well… I feel like buying a round of beer for the bar." "I've never done it so how would I go about doing something like that".

The bell quickly rang again and the announcement was declared for all.

"THIS FINE GENTLEMAN WOULD LIKE TO BUY A ROUND OF OLD STYLE FOR THE BAR".

He pointed at me then followed "OLD STYLE ONLY, YOU'LL HAVE TO PAY FOR THE FRUITY DRINKS AND REAL BEER YOURSELF."

15 minutes later, after the mad rush to the bar had ended of course, strange people kept coming up to me thanking me for my hospitality. One even said "Thanks man, that was really cool, nobody's ever done that before." I felt like a big shooter and it only cost me $5.30. I mean, where else can you make 50 people happy for less than six bucks? I gave the bartender $10 partly because nobody was going to tip him on a ten cent beer but mostly because it was a load of work.

The night rolled on and we drank, we laughed, we tried to remember what 20 year-old females really looked like under all those layers of plaid. I smiled to myself and realized that I had been away from home too long. The evening was very nice overall but it was a long day and it was finally time to get back to my dumpy little hotel and call my wife before sleeping alone, again.

Religious Confrontations

I liked the people in Wisconsin except for the occasional militant bible-thumper, which I seemed to run into from time to time. The next morning I stopped at a local food store for some office/snack supplies and was approached by a young man in the check out line. I must have some sort of look that says "come talk to me about religion, I'm morally depraved and need your help."

"Isn't it a glorious day" said the strange man with wide smile.

You tend to be obligated to answer that question politely by saying "Yes it sure is" then the flood gates were opened.

"It's a day only God can make don't you agree?"

"Yes it is beautiful", "now leave me alone" unfortunately the second thought was unspoken.

"Have you been saved?" "Have you accepted Jesus as your own personal savior?"

"No, I'm Catholic."

"Oh I see" he said in a disappointed tone.

"Only those that have been saved will make it into the kingdom of heaven".

Now I've heard this, and studied this all before.

Coming from a strict Catholic background I was trained as an alter boy and attended private parochial schools for 12 years, I think because my parents felt that it was the best training for their two undisciplined little boys. (We were kind of undisciplined after all). Mass was mandatory at least twice weekly while in grade school, and alter boy duties had us serving multiple masses on the weekends. Perhaps you've heard of our grade school, "Our Lady of Corporal Punishment".

At one point prior to starting high school I actually thought I wanted to become a priest and went off to the seminary for high school orientation in Wisconsin. It was this thing I had about wanting to "help people". Ah yes, another connection to Wisconsin. During a rare free-time break, I accidentally happened across two students engaged in a lip lock and I made up my mind then that this was not the environment for me. I was pretty sure at the age of 14, that I liked girls, not boys.

Ultimately I attended an all-male Catholic high school with daily theology classes, weekly mass where all classes were taught by priests or brothers. (Brothers are in reality former football players in brown robes who are considering the priesthood).

We learned not to piss off the brothers. The point is, we studied theology and the prophets of history daily. It's a subject I could argue with someone about for days and the quickest way I know of to make enemies, but I'm way off topic again.

Back to Wisconsin and the "Food Mart", I smiled at the man who had just informed me I was going to rot in hell and said "excuse me", grabbed my groceries and left. It is simply something I can't stand. To have someone tell me what I can and can't believe or that I'm wrong because I may not agree with a specific interpretation of a book. Worse yet, the challenge and the confrontation came unsolicited. If you study different types of religion at all, you'll find that disagreements about the "interpretation of text" is exactly what starts wars. There I go again, off on another tangent.

The people in Lacrosse have a strong moral background and much of that comes from religious belief and a conservative population. It isn't a bad thing really and it generally makes them a very warm and friendly people. The folks on the project team were terrific and the month I was there flew by. I enjoyed all my time in this little Midwestern town except for one little crusade launched at me on a Tuesday morning at the food mart. He was lucky I wasn't in the mood to discuss St. Peter versus Mohammed or the military ramifications of claiming you're a Protestant in the wrong section of Belfast.

Chapter Six – Manufacturing and Masochism

My cell phone began to ring at 7:30 on a Monday morning.

"Hello, Janice?" "No, I'm back in the Detroit office catching up on some expenses I need to get in before I run out of cash". "Ha! try 6 weeks behind."

I always seemed to wait until the last minute. Janice started to tell me about a local assignment in the Detroit area.

"Its for a large manufacturer in Detroit and they're doing a full suite of Financial products globally."

"Its an awesome technical opportunity for you and a chance for you to stay near the wife and kids you know…"

She always felt like selling the idea I noticed. Janice had apparently liked my work enough on the last couple of accounts to ask for me again and Mike back in Chicago thought it was a good idea too. It was a done deal before Janice ever gave me a call.

It was a big pre-sales installation of our product and based on information in my first day of orientation, would be another test to maintain sanity for all those on the project. Pre-sales meant that PeopleSoft's sales guys would be running around everywhere doing the "build the relationship" thing while we grunted it out in the trenches making customizations to the software and prototyping it as promised. The sales people were tense and made it a point to tell us over and over, "You guys better not screw this up." Now there's incentive for you.

Over the course of the next couple of days I met some pretty incredible people from my own company. This was the first chance I've had to work with another company employee, except for Janice of course. These were all consultants for PeopleSoft but with different skill sets. We had been chosen specifically for the blend of skills because we would all be working to create a prototype that covered General Ledger, Accounts Receivable, Accounts Payable, and Asset Management.

There were 7 of us on the core team and we became fast and furious friends. It was amazing to me that 7 people with such varied backgrounds and ages could be this close but as the weeks wore on and the work continued, we became fast friends. If one person started to have trouble, another would pitch in with a second pair of eyes to look into the problem. If another had a success we all shared in it.

Janice turned into a mother hen looking over the work we were doing, but also making sure we stuck together and supported each other. The occasional dinners out were especially entertaining and we seemed to gravitate towards a restaurant called the "Big Fish". Bob McLaughlin would always order a single malt scotch before dinner and preferred to stay away from brands made in the "peat regions of Scotland".

Bob had turned into a good friend quickly as we used to share stories of bad former employers and poorly run projects. He usually won the bragging rights over the discussion being 17 years my senior and having more chances to work for "assholes at shitty companies". He was truly the undisputed expert in global financials having more experience there than anyone else at our company. I kind of liked the way he laughed. It started as a high-pitched whoop and simmered down to a low series of laughing exhales that trailed off into a quiet smile. Bob was simply one of the most charismatic and loveable mentors I ever had. We affectionately called him "grandpaw". I think he liked that.

Dinner that night was well timed after a long week of customer discussions, arguments and meetings to settle arguments. We were ready to vent frustration, drink heavily and try one of the house specialties, Broiled Salmon served on a wooden plank. I went against the grain and ordered a steak and baked potato, mostly because I never get a chance to have steak although it did seem a bit strange at a seafood restaurant.

The conversation was lively as usual and typical when you assemble a group of similarly educated and field experienced consultants. "Mom" as we affectionately called Janice, was seated at the head of the table with Bob seated at the other end. The rest of us were spread out across this large rectangular table and there was ample room for hand and arm gestures, which fit our collective persona well, at least this evening. Each would take turns telling a story of an encounter with the customer that week told within a framework that suggested that certain people there were "intellectually challenged".

Fred went on to tell about the Accounts Payable product demo followed by a business requirements discovery session.

"We did the entire product demo, you know features/functions, the whole nine yards, and suggested ways that they could implement our module with very few mods".

"Then we did the requirements discovery session and they started talking about all the ways they could re-write the module to make it work just like the way it currently works on the mainframe."

The customer had apparently discounted much of what was offered at the product demo session by saying in summary. "Well, let me tell you about the way we're used to doing things here."

Fred continued. "I about jumped out of my socks". "So I challenged the one guy Tony, as to why they bought a best-in breed software solution if they wanted things to work the old way."

"I told him", In frustration Fred started to point fingers with outstretched arms, "Why do you think your senior management selected our software if they didn't want to change the old business processes?"

Fred was getting worked up about it all over again. Bob, held out his palm towards Fred.

"Now you know these guys are a bit slow, and they've been doing things the same way for 15 years." "Change will be difficult, and it will take time." "Big ships turn slow."

Bob always had a southern, common sense way of bringing everything into perspective. Fred nodded, sat back in his seat calmly and picked up his glass of beer.

Later it was my turn to add to the stories of the week and I chose to talk about various strange encounters with the customer's management team, but one in particular. The technical project lead 'George' had pulled me into his office to talk about batch performance guarantees that our sales team had placed into the contract. George produced the benchmark white papers showing high volume transaction throughput numbers on an unmodified database.

"What?" was my first non-professional, in-shock response. You see, if you understand anything about hardware and software architecture, you should never offer system performance guarantees on your software if your customer intends to modify it. You can't compare published performance benchmark statistics on "vanilla" software to the actual performance of a customer's heavily modified system. It is "Apples compared to Oranges". Our sales team didn't think to ask anyone if this contract language would be a potential problem.

"The customer showed me the portion of the contract that guarantees that their modified solution would perform the same way as it did in the non-modified (vanilla) version."

"I don't know how we're going to get these systems to perform at vanilla throughput levels without extensive tuning… and even then I can't guarantee the same results."

"The sales team just made us responsible for contract language that we have to deliver to, or we'll be in breach!"

I guess I was starting to flail my arms a bit, the people to the right and left of me were moving their chairs further away. I like to think of it as charm.

There was a collective inhale at the table, and a couple of heads turned to look at me while everyone else turned to look at "Mom". She asked for a copy of the software contract, which was entirely different from the services contract we were performing under. I promised to get it to her.

A couple of months later, knowing we had to tune the modified system to get it to perform similar to published benchmarks, we set up a benchmark activity of our own. We were now ready to performance test our prototype at a hardware partner's facility in California for newly released PeopleSoft Financials version 5.

Most of the team made it out west for some part of the testing but one week in particular stood out among the others. It was a hard week of consecutive15+ hour days as we loaded the software and set up the server clusters for both batch and online performance testing. Everything had to be just right in order to glean any good information out of the benchmarking efforts.

We would refresh the databases, load transaction data, apply SQL and index tuning that we wanted to try then run the test again. The iterations went on through day and night.

Because this was an acceptance testing effort for the customer, comparing our results against the contract, they came along with us to validate that the tuning we applied to the database and servers actually worked. There was no "taking our word for it" from these guys. "They couldn't even pass a greased BB through their ass if they wanted to" as Fred used to say.

The Bondage-A-Go-Go

On Wednesday of the first week, the small group of customer folk traveling with us started talking about the need to go somewhere and blow off a little steam. Sounded like a good idea to me, as we seemed to be stuck in a glass fishbowl of a server room for 2/3 of a day. Our hardware vendor on-site representative or "babysitter" asked us if we ever heard of a place called the "Bondage-A-GO-GO". I'm pretty sure the collective answer among us conservative Midwestern folks was something like "uh…no".

He explained that it was a once a week event at this dive bar in the warehouse section of San Francisco. Sound good so far? He explained that you didn't have to dress weird or be weird for that matter, but it was a great place to go and "people watch" in another one of San Francisco's alternative lifestyle communities.

We visited the website, "yes Toto, there was a website", and we learned the next weekly event was planned for this evening. The site suggested that we at least dress in jeans and something black so we didn't stick out like sore thumbs. We of course, interpreted that in the most conservative way possible.

Now I'm not sure what possessed the customer group to take an interest in this whole idea, but they absolutely wanted us to take them there and see just how weird it could get in San Francisco. The other two were short haired family men with at least three kids each and a wife of 10+ years. They all looked at each other and said "Lets do it!" I guess I had to be the entertainment director for the evening. Julie McCoy was always my favorite character on Love Boat anyway.

With time to stop at a local department store, we all had run in to grab our mandatory black tee shirts. We changed, piled into our respective cars and headed into the manufacturing district of San Francisco. We met up with a friend of mine Alan, who worked locally near San Francisco and I dragged him into the fray. He was another PeopleSoft employee and assigned to the pre-sales account in a tech support role and the same guy who had the futon in his office in Pleasanton. Misery loves company and I didn't want to be the only one entertaining this customer at a place like this.

We found a parking lot across the street from this dump of a building called the Trocadero Club and saw a large line of people waiting to get into the place. A large line of people dressed in black leather. A large line of people wearing face masks, handcuffs, chains, capes, boots and all with various and sundry body parts pierced with (insert the name of your favorite metal alloy here).

Approaching this throng of upstanding citizens in our faded jeans, black tee shirts, and buster brown haircuts and it was obvious to all that we just didn't belong. The website said to "Wear something black", but that took on a whole new meaning when we saw the group waiting to get in. We took our place at the end of the line and engaged in quiet conversation, partly to satisfy the customer group who were actually having a good time, and partly to not piss off the rest of the crowd since we were out numbered 150 to 6.

As we stood there, the others in line took note of our lack of costume, but ultimately didn't cause us any trouble. Part of the entertainment was just waiting in line. From behind us approached a tall black haired woman, completely covered in a black leather cape. She was leading her date by a chain as he was bound from head to toe, face completely covered, and feet shackled so he could only take 8 inch steps. He could not see, barely breath and hardly walk as he was paraded in front of the crowd. At first I didn't get it, then it struck me (not literally), this guy was into humiliation. "I get it now! I don't want to experience it, but I get it." You know what I mean.

The doors to the club opened and we eventually got inside. From there all bets were off as to how much visual stimulation you could stand for one evening. Here you could see it all and in conflict to what the website claimed, there was plenty of nudity to go around. I guess they had to make some sort of disclaimer.

There were beautiful people and hideous people covering the dance floor in various stages of undress. A huge projection screen played music videos and the sound system was truly top notch Bose equipment that made your intestines grumble. The most enjoyable surprise of the evening was finding out that the beers were only $2.00, and the lines to the bar were never long. It was clear that everyone else was there to look not drink.

I grabbed Alan a beer but when I turned to hand it to him, he had moved to my left side and appeared to be hiding behind me. He explained that a very large man dressed head to toe in leather was looking our way and blowing us kisses. My first response was "Dude, no way"... but there he was, all 6 foot 10 of him looking a bit too fondly at us. Time to walk away. "Faster damn it".

We found the rest of our party over by the pool table, which no one was playing pool at. You see, you simply couldn't. It was covered in leather collars, and chains, and whips and other implements of dominance, and everything was of course "On Sale". How does one know if $8.50 is a good sale price for studded collar?

We noticed a bunch of people were climbing up a set of stairs at the back of the club so we, like typical adventurers or lost souls if you like, followed to see what we could see. You see at the Bondage-A-GO-GO, the second floor is reserved for exhibitionism. Our customer throng led the way.

There was much more to see upstairs. Things got weirder, if that's a word. There was a central area chained off where couples, at least I hope they were couples, took turns beating each other while the whole crowd watched. There were various implements in use from whips, to chains and even some electrical device that directed static electric shock towards the recipient.

This rather large woman whose wrists were chained to the ceiling just stood there blindfolded while her thin little husband directed static electricity shocks from a rabbit fur mitten towards various parts of her body. She would react to each shock with a jerk of her body followed by loud moans that I'm guessing were for the audiences benefit.

It was dark enough to see the sparks jump off the mitten, and illuminated enough to see other people's reactions. His fur mitten was directed at innocent parts of her body first, but then moved down to other more personal parts in the end. Our group stood there speechless and motionless. I think my mouth was actually open in disbelief. It was strangely erotic and absolutely disgusting at the same time. You felt like you wanted to look away, but simply couldn't. It's like watching something horrible on the news or gawking at an accident on the highway. No one really wants to see it, but you're drawn to the reality of it just the same.

On the other side of the room, there was a line, a relatively long line of people who were waiting to be beaten. At the end of the line was a fenced-in area with a pommel horse, handcuffs on the wall, ropes and chains hanging from the ceiling and one wall adorned with various whips and paddles. In the center of the fenced in area was a young man with long dark hair, dressed in black whose job it was to beat you, that is if you paid him the $20 to do it. We all had guessed after watching the center fenced in area for a while that this guy really, really liked his job. He took such great pride in his work.

The next paying customer was a young male wearing a full leather mask who chose to approach the wall. He faced the carpeted wall with outstretched palms, spread eagle, leaning forward as if being searched by a police officer. The young disciplinarian grabbed a ' Cat O' Nine Tails' (we had to look this up on the internet) and a strange quiet slowly enveloped the crowd. With a eerie silence he swung the whip as hard as he could and a loud 'pop' could be heard as the leather hit the victim's back. It left immediate and significant red marks on the young man's shoulders. The next swing caught the lower back with even more severity, and I could feel my stomach turn.

The young man on the wall would wrench his body in pain, but his position never changed on the wall. Despite not being chained or tied to the wall in any way, he continued to take the beating until his knees buckled. I glanced over to see the reaction of the customer group we were <ahem> entertaining, and all seemed to be as shocked as I was. The young man's back was covered in horizontal red lines, soon to be long lasting bruises. With a collective group nod and no spoken words, we moved towards the back staircase and found our way to the lower level.

We didn't last long after that. Beers were consumed and nobody was interested in ordering more. We had collectively decided to leave and let it all soak in while we made our way back to the valley. It was quiet for the longest time in the car, then the laughter started as we pointed out what each other's reactions were during the evening. "All too different, too strange, definite sensory overload."

I'm sure that we, in our clean black tee shirts and "Buster Brown" haircuts were the subjects of conversation and ridicule ourselves. It was clear that we didn't belong, didn't fit in, and didn't run with that crowd. It could be a discussion of mainstream versus alternative lifestyle, or religion, or even of an American's right to do as he/she pleases. Whatever your perspective, it was too much for me and everyone else in the group. We needed to get out.

It was the major topic of discussion for weeks after the incident. Each accused the other of liking the club a bit too much. We even bought Janice a riding crop from the pool table before we left because she had missed out on the experience. Janice would later learn to treasure that thing too, producing it later on the project when she wanted to kid us about "cracking the whip". "Put it down Janice, put it down".

Job Offers

I had developed some sort of reputation by then among some of the technical folks in Pleasanton. Not so much as a hard core tuning or database guy, but as a generalist across different parts of the architecture who could also manage little projects.

One of my contacts in the corporate offices had arranged an impromptu interview with the then head of the Performance and Benchmark team which was an off-shoot of the PeopleTools group. I would be responsible for certain benchmark efforts, whatever they were but would also have access to virtually limitless resources from our hardware partners. It was very tempting.

While I was still in California, I seriously considered the transition to a more internal role with the company, but needed to talk at length to Cheryl first. The kids were still small and moving in 1996 would have required the least amount of impact to them, being able to build friendships and get used to new schools quickly. Then there was always the cost of living differential to consider. There was some discussion of a pay increase with the job offer, but it would not have been significant. I had enjoyed a premium for being on the road after all, and this was truly a no-travel job.

Another co-worker and recent acquaintance I met from Brazil was also considering a move to the corporate office from his home in Atlanta. We had met in the cafeteria as a small group of technical people gathered quite by accident at a large table in the corner.

We introduced ourselves to each other and as usual at PeopleSoft, everyone immediately got along with each other. I wondered then if there was some sort of personality profiling we had secretly undergone as part of our interview process. It seemed that everyone I had met at PeopleSoft was pleasant and unassuming, with a subtle sense of humor to round out the profile. Lunches with complete strangers in Pleasanton, turned out to always be a pleasant experience. Hmmm. Pleasanton... pleasant. Aptly named, this town.

Since Manuel and I were both considering the big move, we started to look at houses in the area after the work day was done. On one particular day we were able to leave early and based on some advertising in the local paper, we headed out to some new housing development near "Discovery Bay". The trip wasn't too bad we thought, both having come from backgrounds where an hour commute was common, but this particular subdivision was in the middle of nowhere.

The homes were architecturally interesting, with stucco exterior and wide open floor plans to make them seem larger than they were. Some of the drawings available in the model showed square footages between 1300 and 1900 on a slab and lot sizes about 50 by 70 feet.

The sales agent in the model had finished with an older couple at his desk and saw us by the literature. He walked up to us with a big smile and an effeminate voice and declared

"So you two are looking for a home, how wonderful".

"Separately", Manuel replied quickly.

"Yes, my wife would like the floor plan" I added.

"Oh, I see" said the agent as if disappointed.

This was a mystery to me, since he just went from potentially selling one home to selling two. It was talk, talk, talk, for about 30 minutes before we could get down to prices. 1300 square feet, on the fringes of suburban development, on a small landscaped lot came in for just under $300,000. I quickly did the math in my head, knowing that any pay increase to move out here would be minimal.

Back in the suburbs of Detroit in upscale neighborhoods, I could own twice the house, twice the lot size, with a full basement for half the money. For us to move out here with three kids meant finding an even smaller (or cheaper) condo.

Cheryl and I talked at length about the possibility of the move and how we could make things work, or not. Family was a major consideration as literally everyone we knew lived in Michigan.

We pulled out a spreadsheet and calculated potential expenses, house payments, whatever we could estimate and the math wasn't working in favor of the move. In the end we figured out that we could do it, but the intangible factors surrounding the kids and our extended family tipped the scales against moving to California.

Chapter Seven – Home and Away

The California travel came to an end but there was still some clean up to do at the customer site in Detroit. The work continued for a few weeks after the West Coast visit while we collected our victories and our defeats suffered during the endless benchmarking in California. There was plenty of documentation to do and everyone was busy wrapping up our "proof of concept" for the customer.

It wasn't just technology testing and performance tuning. Our product consultants worked extremely hard to customize and configure the software to work the way the customer wanted it to. It was a Herculean effort requiring many evenings, weekends, caffeine and bad pizza to achieve the proposed result. In the end, PeopleSoft won the business and started to become a major player in the global ERP market.

Back home, things were just OK. Though I was considered to be "local", by working in Detroit. I was still an hour away from home in zero traffic. Factor in a 10 to 12+ hour work-day and I was out of the house almost 16 out of 24 hours. The kids were asleep and so was Cheryl when I returned most days, and it might as well have been the same as an away assignment. No one at the Wortham house was happy with the situation, and Cheryl learned that even when I was "local", she couldn't count on me to help out with things at home.

The proof of concept effort was completed and Bob made his way out to Chicago to start up with another financial implementation for an insurance company. I went back to our Detroit office actually located in Southfield Michigan to do a little administrative work and some research into some of the new technology our Tools group was building. Most of the week was uneventful and I had a chance to actually get home for dinner and extended visits with my kids. "Visits".

Thinking about it now, that seems to be how I viewed my encounters with my own family rather than talking about it in warmer terms. I should be looking forward to "going home", not planning a "visit" with my kids. After only a year and a half of travel, I was growing apart from my entire family simply because I wasn't there.

Our sales force was starting to target larger customers on the heals of some successes in the field. It was a simple proposition for them. Bigger companies meant bigger sales and therefore bigger commissions. I suspect this is about as deep as most sales people can go in terms of understanding causal relationships. For them it was a money thing, for us it was a "fix it after they sell it" proposition.

Don't get me wrong. I like most of our sales people. They bring the customers in the door and "new customers" was a metric that Wall Street looked for. But there were those who if tied to cinderblocks and tossed to the bottom of Lake Michigan, would make me like them all the more. Yes I know. They bring in the initial dollars and the contract into the company. Dealing with the actual implementation was my job. I got into an argument one day with a sales guy in the Southfield office after they started "bashing" one of our consultants at a customer site.

"Wortham"

All sales people seem to like using your last name when addressing you as a means to exert some sort of dominance when they're pissed off.

"What"

"What do you know about this guy….. Tom Fredricks"

"I know he's solid in operating systems, knows the database engine, and understands installation procedures pretty well…. Why?".

"Well your boy is fucking things up at my customer and it needs to get fixed now"

"OK". <pause for effect> "First of all Tom isn't my boy, and I doubt he's fucked anything up so what seems to be the problem"

Now the truth comes out. Some sales people don't like hearing from a customer after they sell something. For them its off to the next big thing and "don't ever call me again". Of course he was friendly with the customer on the phone but barged immediately out of his office to look for someone to dump the problem on. That would be me, if you're keeping score.

With a product as technically flexible as ours was, it was difficult to configure and implement if your business requirements called for lots of modifications. You really need to know the product before messing around with it. In this case the customer took their little 6 million dollar installation and starting messing heavily. This is what I suspected, but Tom eventually confirmed it. I went hunting for my favorite sales person with my findings.

"Randall?" (the friendly sales person)

"Yeah?"

"That customer you got a call from screwed themselves <pause for effect>, without a condom".

"They started playing around with the configuration screens before taking any training. They started testing functionality with a broken configuration, and couldn't understand anything they got as a result of their invalid test."

"Do you want the good news?"

"Yeah" was Randall's only response.

"Tom will fix it for them with backup copies the installer took before he left the site".

"I suppose you would like me to take care of that for you?"

"Yeah". "Uh….. yeah".

Not "thanks", not "I appreciate it", nothing but "yeah". There will always be some sort of disconnect between sales and services and I don't think it matters what industry segment we're speaking of. Sales guys like the thrill of the kill. Its like hunting for girls at bars, laying out the best bullshit they can muster, achieving the conquest, then rushing out the door without leaving a phone number.

The service guys by comparison are the poor girl's best (male) friend from college. They confide in us and tell us their troubles, we try to fix things and make it "all better" but we don't receive any extended benefits from the relationship. In a nutshell, that was the role of PeopleSoft Professional Services in 1997.

After a couple of weeks of boredom and a few assignments I could handle from the office, I got a call from my buddy Bob "grandpaw", out in Chicago. They were just starting the project and struggling with the technical architecture aspects of the project plan. They wanted me to join in a conference call to discuss next steps and a likely assignment in the downtown area. After a couple of calls and a confirmation from my manager, I made plans to head to Chicago for an extended stay.

I was met by Bob the first day and then introduced to our "implementation partner", a well known accounting and consulting firm we'll just call "ENFJ". For those of you who have gone through personality and leadership profiling, you might appreciate the acronym.

They had secured the "prime" assignment by taking over the project management side of the project and supplying most of the consultants. The folks from PeopleSoft made up about 20% of the team and were there purely as product experts.

The problem with that scenario was that you needed product knowledge to do just about anything with the product. It was funny to see a throng of well dressed ENFJ consultants preparing documentation in PowerPoint, while the PeopleSoft folks owned most of the task level work.

Heading to the 40th floor, I met with the three people in charge of the technical architecture. Before I could take off my jacket, sit down and drop off my bag, the questions started coming from all sides.

"How does the client workstation get configured?"

"How come I can't switch from database to database with the same workstation?"

"Why does the database need more table space definitions?"

"Why does PeopleSoft deliver 6 indexes for a table when it only uses 3?"

"Wait a minute there guys" I started to explain "This isn't plug and play software and neither is the database solution that supports it". "We have plenty of time to talk about core architecture as well as database solutions while I'm here".

"Where should we start?"

I opened my laptop to look for a couple of hardware architecture drawings I had prepared for other customers, and we started to go down the "training" road. My time as an architecture expert was being paid for by the customer so that I could train these people from our consulting competitor. This insurance company was paying me to train consultants who were also being paid to be experts. You just have to love how that works.

About midpoint in the morning the adrenaline that normally accompanies the initial meeting with a new customer wore off enough for me to notice that I really had to find the restroom. Bad. I found a breaking point in the conversation and told the small work-room filled with people that I would "be back in a second".

As I moved towards the door, Roger, the youngest of the "experts" stopped me to ask me another technical question. I turned, replied, then walked towards the door again. Roger followed me into the mocha colored tile hallway with chocolate brown painted cinderblock walls to ask something else. I looked at the floor and the color alone seemed to influence my body towards another immediate need. I answered the question then moved down the hall with the sound of Roger's voice echoing off the barren block walls. I placed a hand on the bathroom door and the questions kept coming. I stopped, turned and as professionally as I could said…

"I really need to go". "I'll be back to the room in just a second, I promise"

Roger, was taller than average and skinny with a dark Mediterranean look about him and a full beard covering his face. The type "A" personality description fit Roger well as he jittered all through the meeting and picked at his own fingers while speaking. No more caffeine for you Roger, at least until the Ritalin kicks in.

With my last attempt to call for a little privacy, Roger chose to follow me into the bathroom anyway.

"Just one more really quick question". Roger continued.

I entered the stall, closed and locked the door and proceeded to enjoy my almost 45 seconds of privacy. Well, sort of. Roger was expecting an answer to his last question. I hope he enjoyed the sounds of splashing and flushing that accompanied us in the bathroom.

Later that evening I met up with "grandpaw" and the other PeopleSoft gang already on assignment there. We all agreed that there were many challenges on this project, but that it was no different from other large implementations we had been on. I shared my bathroom story and it seemed to be unique in terms of consulting experiences. It was a first for me as well. We had dinner and a few laughs and I instantly felt comfortable.

After dinner Bob told me that it would be likely that I would be there for quite a while and that I should look into getting an apartment as a cheaper long-term alternative. After two weeks in the Palmer House Hilton, I was able to check into the corporate apartments down the street on Wabash. It was right next to the IBM building and a street away from the House of Blues and the Chicago River. It was an awesome location if you were there to explore Chicago, but as I found out my first night, not too quiet.

I had a small, furnished studio apartment on the 41st floor which faced the IBM building. All walls on one side of the apartment were actually windows covered in eggshell colored mini-blinds. "Attractive enough" I thought until I noticed the intense light coming through them from the IBM building. It was then that I determined that IBM must be doing pretty well financially, because the lights seemed to stay on all night long, unless of course the cleaning crew was partying every night on every floor.

With extra pillows stacked up on one side of the bed, a large enough shadow could be cast to provide a "dark spot" for my head. And when I wanted to read at 3 AM, I could do it by the light of IBM without flipping a switch in my room. 3AM was a key time of night because it was at this time, typically, that the bars closed and the arguments between drunks started in the streets.

To this day I can't explain why double pane glass on the 41st floor can let in so much noise. I could hear bottles being thrown into dumpsters, arguments in the streets, and the endless parade of fire trucks with full sirens pouring out onto empty streets.

We found later that after a long brain-beating day, beer helped knock us out for the night so a stop was planned on the way home most every evening. Burgers, beers, a 10 block walk, a hot shower and a pillow. It was easy to crash after that.

The months dragged on and there was always more work to do than there was time to do it. The team of people we had on the ground were overworked as were the customer's employees assigned to the project. Tempers flared in the fall of the year when one of our consultants said a few things in a conference room that one of the customer managers didn't like.

Michael from our consulting group was not a large man standing about 5 foot 8 and all of 170 pounds soaking wet. He was well spoken and well dressed and very knowledgeable about reporting tools and customer best practices. In comparison, the customer manager in question was ominous. Kevin was a dark figure of a man standing about 6 foot 4 and close to 300 pounds. He dressed in rumpled business-casual clothes and dispensed four-letter words with a regularity that confirmed his working class background.

During this meeting, Kevin objected to some of what Michael was recommending. Rather than have an open discussion about the disagreement, the conversation quickly moved to a shouting match with plenty of colorful language and accusative finger pointing. After the meeting was over, Kevin cornered Michael in the hallway next to the elevator and with a finger firmly poking into Michael's chest said

"I should take you out to the alley and beat the shit out of you".

Michael found me at my desk shortly after and grabbed my arm gently to get my attention.

"I need to talk to you about something that just happened"

"What", "I mean sure" "What happened?"

Michael looked composed enough externally, but his voice was racing a bit.

"I was just threatened with bodily harm"

"Where, outside on the street?"

"No here in the hallway, next to the elevator"

"What?"

I questioned him with a slightly increased volume mostly because it bothered me that any of the good people in our group would fall victim to something like this.

"Kevin just threatened me next to the elevator"

Michael went on to tell me the story about the meeting, including the build up to the shouting match. Most of the shouting was done between the customer's managers but it was normal in this corporate culture to let meetings get out of hand like that. Four letter words and phrases like "fuck that" and "your full of shit" were surprisingly normal in meetings here. He explained that Kevin took exception to recommendations Michael was making when he had a different understanding before the meeting started.

"So what do I do?" "I mean I think I'm ready to head to the airport right now".

"Well, I think he needs to be confronted with his manager in the room".

"Basically we need to get the issue brought out in the open and Kevin's behavior needs to be reviewed".

"It should be a clear HR violation here, and there is no dispute that you now feel uncomfortable in this workplace". "It wouldn't matter if you're a consultant or a permanent employee....".

Michael thought about what was said for a minute. "Well.... I'm not trying to get anyone fired over this"

"At this point all that matters is that it happened and you're uncomfortable with the situation". "The behavior is actionable and I'm willing to push the issue to his manager".

"I'm not sure what I want to do now" "I don't want to damage our relationship with the customer, but I know that I don't want to work with this guy anymore". "Maybe I just need to arrange to get out of here when a replacement is available".

I applauded Michael for his concern over customer relations, but if it boils down to an issue over personal safety on the job, no further discussion is needed. We both figured that Kevin was just being the bully that he felt he could afford to be, but it was also true that he did threaten a co-worker. That alone could and should get the man terminated.

"Michael, I'm willing to back you and take this thing as far as you want it to be taken, but regardless of how we address the issue with his management team, I will look to get you out of here".

Chapter Eight – Sunset on Chi-Town

Michael was eventually moved to a local assignment in Denver where a large insurance company was struggling with reporting issues. Kevin ultimately apologized to Michael, maybe realizing how far he had stepped over the line. Michael never chose to file a complaint against Kevin, but to some degree the damage had been done. Everyone on the project team heard about the incident including Kevin's manager. He seemed to come to work with a different demeanor after that. Still ugly and ominous, but it was a subdued form of ominous, if there is such a thing. There was nothing that could be done about the "ugly". Kevin had that in spades.

The work became more stressful as we pushed to get the product closer to production status. Bob was harder to find as he seemed to be locked in closed-door meetings all day, every day. I would run into other people on the team from time to time, but the were all as busy as I was just getting through the task work that needed to be done. The ENFJ people were doing a great job smooth-talking the customer and doing boxes of documentation in PowerPoint.

A smaller subset of our group started to become close, sort of like a second set of cousins. This is a phenomenon I've experienced at other customer sites and at other jobs I've had. When you spend 18 hours a day with people in stressful situations, leaning on each other for support and stress relief, you become like brothers and sisters. Sort of, anyway. There is definitely no sharing of underwear or anything like that.

After our last big effort with the Detroit manufacturing company, Bob, Jason and I became tight, and to this day I feel like I would do just about anything for those guys. Here too in Chicago. I had continued working with Bob because we trusted each other. We were the two oldest guys on the project so by default we became like older brothers to the other young ones on the team. In Bob's case, he was more like "dad" than anything else. We still called him "grandpaw" every now and then and he would just laugh.

In Chicago, Bob, Jason and I connected with a new family. Michael, Katie, and Karen although there were about 6 others as well. We started doing regular dinners and bar visits after long nasty days at the office. Every now and then, one of the remainder on the team would ask to come along with us, and it was clear that our little group had grown apart from the whole of the team. Bob had commented that we needed to do a weekly team dinner just to keep the entire group together. Not that we wanted our little extended family to be exclusive, it just sort of happened that way.

Our personalities and sense of humor, educational backgrounds and experiences were so similar that we would have become best friends regardless of circumstances and logistics. It was this customer and this city that brought us together but we will remain friends I expect until the day we die.

Summer came and passed as quickly as Milwaukee beer passes through the body on warm summer nights. There was barely enough time to do group dinners or enjoy any of the downtown area. Despite the crazy work schedule, we did manage to do one or two evenings out.

Body Shots and The House of Blues

Katie was a bit of a wild child. She had adopted California as her home and carried the look as well as the lifestyle of a true California girl. One night Katie had organized a little outing for about 9 of us at the House of Blues, not necessarily for a concert but for the food and the atmosphere. We were given a long table at the back right of the restaurant, away from the center bar area. The food and the alcohol started to flow. It had been a bad week full of work and re-work and problem resolution.

Since Thursday was the last night in town for us before we all flew home for the weekend, Katie figured that this was a good evening to get trashed. Sometime after 11:00, Katie asked Bob if he knew what a "body shot" was. Bob looked confused, and so did the rest of us Midwestern republican males. The demonstration began shortly after.

Bob was instructed to hold a lemon wedge in his mouth, lemon-section side out. Katie next gave Bob a liberal licking on the side of his neck followed by a liberal shaking of some salt on the wet spot, as it were. She licked the salt, drank the shot of tequila and bit the lemon wedge that was clinched between Bob's teeth. I had simply never seen anything like it.

"Now do me" Katie said.

Bob looked her in the eye and said "I better not".

"I'm an old man after all and you don't need me falling dead of a heart attack right here in this restaurant".

"Worse yet, I can't imagine having to call my wife and explain why I was in the hospital or what I was doing to wind up there". I was on the same page as Bob.

"Who then?" she asked.

Hands went up around the table from the younger males looking as if they all were in grade school eager to answer a question from the teacher. One young lad who was in training as a "shadow" volunteered to assist her only Katie turned up the heat a bit. The shadow consultant from Florida named Mel, licked and salted Katie, but Katie decided to nestle the shot glass in her bosom.

The blouse she wore conveniently allowed open access to the shot glass begging the question, "Had she planned this all along?" Mel paused and Katie just smiled. "C'mon" she said motioning her hands towards her chest. Eventually, Mel took the glass with his teeth and followed up by taking the wedge from her mouth.

"Ahh" I thought. "I love this company."

The salting and licking and drinking continued around the table as the boys traded shots with the girls, in public at the House of Blues. Nobody could bring themselves to removing the shot glass with their teeth like Mel, but we had fun anyway. Most of us were married, and most of us felt a little uncomfortable with what was going on but we sort of participated anyway. Not to the extent that the single youngsters did, but we were covered in salt just the same. We were all stupid males, this is what stupid males do.

About 1:00 AM, challenges were made between two of the younger males in the pack. One of the ladies was asked to lay back on a table-top for the challenge, and her stomach was used as the salt platform. We were attracting a crowd at this point. The two males took turns salting and licking while the rest of us looked on in disbelief with mouths open. Katie laughed and enjoyed watching the demons she had helped to unleash. It was an effective way for everyone to blow off a little steam, and a perfect way to get thrown out of a bar.

We were trying to get the bartenders attention after 2 and a half hours of solid tequila drinking, but he kept ignoring us. A few seconds after that, the manager showed up at our table to ask us to leave. I guess the show was a bit too much for the House of Blues. On the way out, people were asking us where we worked and what they needed to do to join our company. It was truly memorable.

Friday morning I woke up feeling twisted and sick. My arm was wrenched behind my back and my legs were folded with one up under the other and my foot stuck under the weight of my ass. I must have looked like Marty McFly all twisted in bed with clothes still on. Apparently the only thing I had taken off was my shoes. I unfolded myself and tried to regain feeling and circulation in my foot and arm. I glanced at the clock to see it was 10:45. I'm supposed to be at work at 7:00.

I got out of bed and stood up only to feel something like sand running down my chest and back, past my underwear, down my pants and into my socks. The salt was everywhere. I pulled off my sticky clothes which stunk beyond belief, and limped my way into the shower with a hibernating foot.

Hot water pounded my face in the shower but did not help the hangover and I was angry with myself for missing two meetings that I had scheduled this morning. The long painful walk continued past the Chicago River and down Wabash.

15 minutes later I arrived on site to make my apologies for being late. Everyone else was there including Bob who gave me a smile and pointed to his watch and said "I was here at 8:00". Making my way to my desk I realized that I never called my wife the previous evening as is the normal procedure when I get back from dinner at the end of the day. I settled into my desk trying to figure out what say as I punched the area code into the phone…

"Hi, its me"

"What happened to you"

"Well I guess the gang needed to blow off a little steam last night…. So we did"

"You did… what exactly"

"Well it involved a lot of tequila and salt and…. lets just say that it will be a long story I'll save for later tonight when I get back".

"What time is your flight?"

"Its at 4:30 so I should be home by 8:00 at the latest"

"OK, I'll see you then. There are no kid activities tonight so we'll be here".

"See you then"

My head was pounding and I needed food. From past experience the thing that works best with my stomach and with my metabolism is greasy food and Cola. It was almost lunch time when I got off the phone so a small group of us went out to Bennigans on Michigan Avenue for lunch and a big honkin' cheeseburger. I started to feel better after the third bite. The afternoon was tolerable after that and I made my way to O'Hare at 2:30.

The explanation at home turned out to be a disappointing one for Cheryl but I guess she appreciated the truth rather than me not telling her anything. Most guys would have said nothing at all. It was after all mostly harmless fun that I probably should have walked away from, but I figured if "Grandpaw" could hang in there, so could I.

Weeks dragged on into months as we braved the winter in Chicago, trying to get across the bridge on the Chicago River without being blown into the street and into on-coming traffic. The days continued to be long and the hours normally used to off-load stress with our gang were being reduced by the day. It was all we could do to get out of there by 10:00 or 11:00 PM in enough time to grab a carry out order before all the restaurants closed. Some evenings we were limited to pre-packaged 12 hour-old sub sandwiches from the White Hen Pantry on Wabash.

A Consulting Break in San Antonio

The national consulting group had grown to a few hundred people in 1997 and we were still small enough to be able to pick from a number of cities to hold our annual conference. San Antonio was selected and I was asked to provide a training session on "consulting skills" during the breakout sessions. Attendance was as mandatory as they could make it, given the fact most everyone was completely billable, but the conference crossed over a weekend so as not to impact our customer's projects.

My first consulting conference was held in Miami the year before and I remember there weren't more than a hundred people in attendance, or so it seemed. The San Antonio gathering was huge by comparison and the meetings were more formalized. There was plenty of networking time as well, so that we could share ideas and figure out how somebody was able to fix a certain problem where you couldn't. It was a great event.

Ah, but in order for San Antonio to be a PeopleSoft "great event", it had to live up to its growing internal reputation. The "Work hard. Play hard" phrase was about to be tested the next two nights on the Riverwalk. The first night started with a team building exercise where each table of people (10 to a group) formed a scavenger hunt team.

Initially our little team from Chicago sat together but the rules of the hunt forced us to break up into teams of people we didn't know. We were given a limited amount of time and a list of things to collect on the streets of San Antonio. It was a "team building" event because the hunt was designed for interpretations of whatever items were listed.

Our group actually did very well deciphering certain clues and thinking outside of the box, as it were, to complete our list. We weren't the first group to return to the conference room, but we were the winning group on overall points.

I remember being called as a group to the front of the large conference room in front of Dave Duffield, Peggy Taylor, and the rest of the senior management team.

We were about to be presented with our little award when Katie came running from the side of the stage and leapt into me, wrapping her legs around my waist.

I stood there shocked, while everyone looked on in disbelief as I tried to make it a comedic moment by walking away from the stage with her still attached. It was funny after all. Completely unexpected, but funny. Cheryl would have had a few words for me afterwards I'm sure, except I didn't do anything to prompt it. Maybe it was a good thing she wasn't there. There may have been a death to report to the San Antonio police that night. Mine.

The River walk afterwards was a blast. I found the rest of the Chicago gang and we moved with the crowd from bar to bar, wherever the management had a tab open for PeopleSoft. Bob and Katie, Jason, Michael and I found a nice little corner of a beach theme bar and settled in for some conversation and a few jokes.

We still shook our heads about Katie's antics and she laughed along with us just enjoying the evening and waiting for it to get wacky. Eventually the bar top dancing started and Katie was lost in a mix of other males and females shaking booty to whatever was playing in the background.

We moved onto another bar that had been closed down for our private usage and again the tab was open. Somebody came in from the walkway and called for us to look outside, because the North American VP Jim Bozzini, had just been thrown in the river.

"Damn, we missed it". And "Why aren't they at this bar?"

The classes continued the second day, and that evening's event was held at a dude ranch. Pretty cool actually where dinner was all BBQ. We later grabbed beers and watched the Armadillo races or goofed around in one of the other areas of the complex. Bob, Jason and I eventually found a quiet corner in the main hall and started a cigar, which I rarely do but it seemed to fit the mood at the time. We wrapped up the next morning and we would all head for a short respite at home before beginning another week on the road.

Back in Chicago

As we moved through Fall and came into the Christmas break, Katie surprised us again by presenting the four remaining guys on the team with a gift from Victoria's Secret. We all looked at each other across the dinner table from Katie who sat there smiling as if she were the Cheshire Cat. It was a pair of red silk boxer shorts apparently sized perfectly for each of us. There was a pair for me, Bob, Jason and Michael. Katie broke the silence by telling us that the presents were really for our wives and that we should wear them to bed on Christmas Eve.

Jason looked at Michael and said, "I'm not sure I can bring these things home... I mean how would I explain this?" Remembering what I went through after my tequila incident, I offered "The truth is far better than anything you could whip up as an alternative", "or just don't bring them home at all". Katie frowned at that, but I promised to keep mine and offered my thanks for her gesture.

The flight home that week had me bump into the wife of who used to be one of my best friends. John Shane and I had known each other since teenagers or maybe that was just me at 19. I met him at the Sheraton Hotel in Detroit where he was a maintenance engineer and I hired in as a dishwasher.

John at 6 foot 7 and me at 6 foot 6 were the tallest employees in the place and that was enough of a common attribute to eventually bring us together for after work dinner and drinks. Basketball was our common interest at the time and we went to as many Pistons games as we could afford.

Back then they played inside the Pontiac Silverdome in a horrible venue created by using half a football stadium separated by curtains and buffered on one side by portable aluminum stands. You had to be a real fan back then, there simply wasn't any comfort in the seats or any performance on the floor for that matter.

"Donna, shit!" Which of course I didn't mean to say, but I was truly surprised "How the hell are you, what are you doing here?" Another little slip.

"I'm on my way home after some meetings in town". "You're still traveling a lot it looks like?"

"Yeah, the fun never stops". "I've been out here for a few months now and its all I can do to get home and spend some time with my family".

"How is everyone?"

"Good... I think." "I get my nightly update but don't really catch up with Cheryl or the kids until I can get home." "I'm sorry I haven't called you guys, but my weekends seem to be filled with honey-do lists and kid activities" "There's just no time left".

"Yeah, I know how it is" she lied.

"So… here you are at O'Hare", "Are you traveling a lot now?" I asked.

"No, just every now and then and when I do, all sorts of things at home seem to go wrong". "Not big things, but little problems with the kids or things that need to be fixed or teachers that need to be called". It wouldn't be a problem if I was home." "I don't know how you do it".

"I don't do it well, if that's an answer, but I seem to keep doing it". "Anyway, what's up with John these days?" "Home repair, outdoor projects, How is the old boy?".

"Good, he's moved to another hotel where he became the chief engineer and he seems to like the property manager well enough." "No real projects at home, maybe some landscaping outside but nothing big."

"You know, you should call." She said, "He's been talking about heading to another Pistons game when you're in town".

"I know, I need to…. And I will." "You take care and have a safe flight, say Hi to John for me."

And with that I realized that while we were walking, I was just following her as she headed to her gate. I was nowhere near my gate and needed to backtrack to get to my plane on time. "Where was my head".

The remaining weeks in Chicago were busy as hell and there wasn't enough time to go out like we used to. It was all about getting this thing into production and getting the customer live. After a few hiccups and false starts, they did just that. The team started to split off to other customers and other assignments and Bob was pushed off to another global financials implementation for an Asia-based auto manufacturer, and Jason was moved out as well. I stayed behind to manage the team a few weeks more but in the end moved back to other responsibilities in the region.

It was tough to break up the gang, but harder still to realize that we found a way to get through a mentally challenging assignment by relying on each other rather than on our own families. Once things started to calm down and I found that I could actually sleep more than 3 hours in any one time period, I realized that I had spent a ton of time away from my family for the sake of this customer and my job.

Chapter Nine – Building a Technology Practice

It was strange being home and reacquainting myself with the 4 people that lived in this house. It was my house, my wife and my kids, but I had spent so much time away from them that it seemed strange to be there. My other "cousins" in Chicago had grown to become my remote support system, and I only needed to come "home" to sleep and do laundry. It was a difficult time for Cheryl. She had gone back to work to pursue her career with medical office billing and office management, and that meant that the weeks were hard on her as well. Even more so as her regular work-day started early to get the kids ready for school, pre-school or day care. Her day only ended when the kids were fed, bathed and bedded and the house picked up.

We often wondered if this was really the American dream. Two income family, three healthy kids, bigger house, nice cars and it was all supposed to be perfect. The outward appearances covered many internal flaws. Cheryl was getting beat up day after day with the do-it-all-alone work and kid schedule. I was working as many hours, but I had a support system to fall back on in Chicago. Things started to crumble a bit. The kids grades in school were suffering, and the kids started to argue and challenge "Mom" at every turn.

It was true then that our children were spending more time with other adults (day care) than with their own parents. When they got home, it was difficult for them to adopt different house rules, or deal with a mother who was already tired but still needed to get through the normal end of the day. "Do your homework", "Pick up those shoes", "Who left the empty cookie package and milk glass on the counter?"

I remembered back to one fateful night in Chicago after a rare but nice Italian dinner on Ohio street, I decided to call home to check on the events of the day. I was in a relatively good mood that day because we had finished a big piece of the project development work and the customer was actually happy with the result. Back at the apartment after dinner I was lonely and missed them all, and decided to call. After going over the events of the day I made the mistake of talking about dinner... and that was the last thing she wanted to hear. She let me have it....in her own quiet way after her hellish day.

On weekends, Cheryl and I both slept in, then went about the business of doing and fixing things that could only be accomplished on the weekend. The days were filled with work activities only because there wasn't time during the week to accomplish them. I had started finishing the basement and was making multiple trips made to the home center for more drywall or 14-2 romex wire, or some other specialty tool that I now needed to complete the project. Cheryl took on laundry or floral landscaping, or some other cleaning job that she could only get to on a Saturday or Sunday. The kids simply did their own thing in the neighborhood, stopping by for lunch then moving on. Strange kids roamed our house all weekend long following our own children, and on one Saturday I counted more than 12 kids in the house while only three of them were ours. It was a zoo.

There was no real "quality time" with the family, to steal a term from the 90's. The entire family was doing their own thing, and doing it very well. We all seemed to go about the business of finding things to do rather than working together to do things as a family. Tempers often flared, mostly over inane things that would roll off most people.

I would find myself yelling at one kid or all for leaving a dried up drop of milk on the counter. At times we all seemed to avoid each other. Monday morning eventually came around and I would be back on a plane, flying to a destination where I felt I was respected and where people listened to what I had to say.

Things seem to get back to "normal" after I left for the week. Cheryl and I would talk in the evening about school, kid-activity commitments and family events where we had to plan in advance our travel and work schedules to attend. Old friends were left by the wayside mostly because we were never in one place long enough to participate in anything they invited us to. Eventually we just stopped calling each other. Even the best man at my wedding and one of my best friends overall, eventually gave up on me and drifted away.

I found myself being apologetic to other family and friends as I tried to explain exactly what it meant to be a consultant. There really wasn't any control over my own schedule. We often received less than 24 hours notice before we had to board a plane to another strange state or country. It was all part of the "glamour" some of my sisters-in-law thought.

"C'mon" they said in disbelief, "Hotels and restaurants, exotic city destinations and great night life".

"No, not really".

"I spend the day in a conference room often without windows and don't get back to the hotel until it's completely dark and too late to order room service."

"I guess I could go out to a bar, but I have to be at work at 7 AM anyway... so... what's the point?"

The fact that I was simply "never home" had changed the family dynamic and affected relationships inside the family and out. It didn't matter what the reasons or excuses were, I just wasn't around. You could no longer "count on Peter". I think that bothered me the most. I liked being the guy you could count on and my family simply couldn't.

Dave Duffield continued to sew the seeds of our internal corporate philosophy which included the paraphrased mantra "work hard, play hard, clean bathrooms". It was all about having fun at work, and working hard for yourself, your co-workers and most importantly the customer. The bathroom comment had always puzzled me until Dave offered the analogy about how a customer would feel about his software provider if they couldn't even manage the maintenance of their own bathrooms.

I had worked at other companies where there were attempts at defining a similar set of "core values", but they were just that, attempts. The oppressive and dingy work environment continued at each employer where the only core value that remained in the end was… WORK HARD". It truly was more of a family here at PeopleSoft and you would do just about anything for your brothers and sisters here, and of course Uncle Dave.

Give me visionaries and leaders over tacticians and managers any day. I remember seeing Dave and also Baer Tierkel (the head of the Tools group back in 1996) in Pleasanton, California at an internal meeting. It was amazing to hear the vision of business technology, the proposed solutions, and how we were pushing the limitations of desktop and mid size computing platforms in the '90s.

It wasn't only cutting edge technology, it was bleeding edge. We the lowly field organization, had to deal with the brunt of that bleeding edge when implementations and go-lives didn't go as planned. At least Dave and Baer thanked us for our efforts and recognized that it was the field organization that helped to create the client reference.

Since PeopleSoft had begun selling to larger customers, new issues started to arise with the sheer volume of transactions the software had to process. The program logic was sound, but the volume of report activity or on-line maintenance activity would stress even the newest and largest of UNIX based computers.

Right about the time PeopleSoft 7 was released to the public, the regional management teams saw an increased demand for technical resources to help resolve configuration and performance issues at customer sites. There were a few people, me included that were early champions of building such a team of consultants, mostly because we could see the need at every customer site. There were other "purely technical" resources like me in other regions but very few when compared to the product specialty people.

I floated the concept of a technical consulting practice to some of the management people I knew in the Midwest region, and they seemed to buy into the logic. We already had a separate software install group, but nothing in the way of people who could handle large systems configuration, architecture, or tuning. There was demand after all, why not fill the void. Some of course viewed the idea as risky, given the fact that we were supposed to be "product functionality experts" first and foremost, but as the product lines grew and the customer size grew so did the problems with performance and system reliability.

Eventually I wrote up a small business plan for the justification of a pure technical consulting group. I talked with guys I knew in the West and other regions, and started discussions with Mike Maszka the regional consulting VP in the Midwest and Jim Bozzini who was the national consulting VP. Approval for the idea came quickly albeit informally, and suddenly I was a technology consulting practice manager. It was the first such group officially formed in the company.

Recruiting was easy and there were quite a few people looking to specialize in technology rather than HR or Financial software. We built a small team at first, and quickly separated our folks into specialty categories. If database tuning was needed, we had a couple of consultants that specialized in just that. If it was system configuration or report processing, we had experts there as well. Best of all while other consultants billing rates were being negotiated, the technical team's rates remained at a premium. Our profitability margins were always high.

The new practice management job was also supposed to mean "no travel" or limited travel which thrilled my family but there was a difference between the theory and the practice. No pun intended. Most of the other managers were located in Chicago, and so were the meetings.

I found myself flying to Chicago for a couple of days most every week, in addition to the customer work I was finishing up at one company in Downtown Chicago. Back then not only did we have to recruit and manage the people in our groups, but we had to find them work as well. Later in 2001, there would be a full department in each region responsible for staffing the available consultants.

More time was spent at home overall. If I had to travel, it was typically during mid week and I could get back on Thursday night or early Friday morning. It was at least a long weekend with the kids, which is actually more time than I had at home even while working at the bank or insurance company prior to PeopleSoft.

The dual duty of consulting at customer sites and managing the team became overwhelming after a few months. We continued to build the consulting team but it was increasingly difficult to get away from the customer work. When a problem was escalated in the region, very often it was me they would send to mitigate the risk or solve the problem. The customer work was still enjoyable, after all you were there to fix something and there is that sense of satisfaction afterwards. Demands for technical resources however, continued to grow.

Our new regional consulting VP had his performance targets to meet quarter to quarter, and the technical consulting practice was one of the groups identified for growth. It got to a point late in the year where I spent more time interviewing candidates that I did doing anything else. Numerous responsibilities started to fall by the wayside and I simply wasn't doing my job. There were "quality issues" during the interview process as well. It was the responsibility of our recruiting department to screen and schedule candidates based on our hiring targets, but it was up to me to accept or reject them.

I found myself explaining why I didn't feel that certain people were worth hiring, and defending my position. In one case I felt strongly about not hiring one candidate because of poor language skills even though the technical skills were solid. "After all", I told my critics, "we're consultants and we need to lead the customer, guide them, make complex recommendations". "You need to be able to read, speak and write English pretty well". "Even the resume was full of grammatical errors".

My arguments fell on deaf ears. The hiring targets were aggressive and I just wasn't going to win my insignificant and narrow minded argument. While I was on another customer site visit, the resource in question was hired by my VP and placed in my group without my immediate knowledge.

The Technology Practice grew to 15 consultants, three of which I recommended against hiring, but customer homes were found for them all. If there wasn't customer work available, we found internal development or tuning projects for them out in Pleasanton.

It wasn't just the lack of control over hiring my own team, it was the idea of what our technical resources were supposed to be used for. Even in the early years of our practice, I felt like we should be providing a service that our consulting competitors couldn't. Not just focused on programming, or server configuration or database tuning, I thought we should be preaching architecture and technical support infrastructure "best practice". A bit lofty in terms of goals back in 1997, but it was the kind of advice that nobody else was providing to client-server customers back then. We sort of had the corner on the knowledge market there. It was the kind of group I wanted to build.

Selling that idea was the hardest part. Our sales force had a tough enough time competing for other consulting roles on a customer's project. Now try telling their senior management that we wanted to help guide their global architecture strategy using our technical consultants. "We have our own technical people" they would typically say, or "Our prime vendor is providing all the technical people we need".

Still, we had internal support from various disconnected groups inside PeopleSoft. It was our team that was often called upon to figure out issues on-site if a project went bad. Statistically, it was never a project that our people were leading, but another competitor's team that couldn't figure out our product or its technology.

We had developed a management fan club after fixing a few of these problems, but much like the red-headed step child, we always wound up sitting in the background waiting for the next chance at some attention.

Annual performance reviews came around in April and I received my first-ever poor evaluation. I was a little taken aback given the struggle I had in building the team and the fact that I could never shake the responsibility for customer site work. I was told, that although they knew I was doing the customer work in addition to managing the team, it was only the management piece that I was rated on. That was my job title after all. I understood, but I didn't. Technical Consulting practices had sprouted in the three other American Regions and one was just getting started in Canada. Granted I didn't build them, but I helped to make the concept viable, "Didn't I?"

My relationship with the new regional VP was never all that great and I continued to struggle in balancing the never ending requests for customer site visits along with the administrative work for the practice. In the end after a little more than a year of managing the technical team, even getting some positive upward feedback reviews from the people reporting to me, I started looking for another role inside the region.

Chapter Ten - Gunshots and Pork Shanks

It was a hard decision to give away the practice, but it was time to be honest with myself on strengths and weaknesses. When it came to administrative responsibilities versus customer facing responsibilities, there was always a draw towards the interpersonal communication side rather than the "sitting at the desk" side of the job. Building relationships with the team, and working together to help solve problems and "make people happy" seemed to be part of a core need for me. With that the conversations were started with the VP about handing the practice over and as expected, there wasn't much of an argument from him.

It took a couple of months of interviews, but the VP selected two managers to split the technical group in half and build from there. I took a bit of solace from the fact that it took two guys to do my job, but that wasn't really fair. The goal was to grow each group to 15 to 20 people by the end of 1998 but they each started with 8.

PeopleTools 7.5 was being planned for release later in 1998 and the early press on the technology was released to the public. It offered a middleware layer on non-traditional hardware platforms and handled on-line traffic more efficiently. It would become an option now for some of our global customers and we expected the demand for technical guidance on the new technology to blossom.

When some of my family of friends heard that I was back on the market full time, I was asked to take the lead technical position for an international project based in Milwaukee. The goals on the project were far reaching but if successful, would provide for the first global PeopleSoft Supply Chain solution in the world. I took the job.

I tend to have a preference for Midwestern towns it seems, having written about them or expressing my appreciation for them in the past. In many ways your own opinion will be a function of where you grew up. I mean, if you're a born and bred New Yorker, you tend to love the fast paced dynamic that can only be found there. If you're from the left coast, you tend to love it there too, even with the overcrowding and overpricing. An outsider such as myself could not possibly convince you to live anywhere else. We all tend to have a unique affinity, an understanding, or an appreciation for the street, the town and the state where we grew up.

Most folks hear me talk about Milwaukee and then suddenly become uncomfortable in their seats as I start to espouse the fine attributes of this Lake Michigan shoreline city. I mean really, who thinks about restaurants, entertainment, the arts, and great quality of life when their only perspective on Milwaukee came from old episodes of "Laverne and Shirley". Yes, bowling, beer and brats would be the stereotype left over from that show, but not representative of this town.

My first impression of a city is usually based on the airport and the ease of entry into the city. The General Mitchell / Milwaukee International Airport is a small one in comparison to other cities in the Midwest, but courteous just the same. I'm not sure why its deemed to be an "International" airport unless you count the DC-9 service into Toronto on Midwest Airlines. Whatever. I got to know the ladies at the Hertz counter after a few weeks of flying into Milwaukee, and I must say they were perhaps the nicest and most helpful staff from any airport I've visited.

I dropped off a couple of bags of hard candy one day just to say 'thanks' for all their help and it was like I just handed them winning lotto tickets. I gathered that people didn't usually do anything to thank rental car employees. It seems that a big part of their job description is to take abuse rather than accept thanks, but I always appreciated their effort.

A Great Milwaukee Customer

During my time at that particular customer site, we had assembled a great team of people to work on the globalization project. Some nights following a long and stressful day at the office, we chose to head back to the Residence Inn just north of Milwaukee for some "down time". There we would organize our little group, decide on an easy menu and head off to the market for supplies to make dinner. The hotel or apartment becomes your refuge and your co-workers now turned close friends, become another support system. We understood each other, we supported each other, we laughed together, we cried together, fights would start about who ate the last cookie.

Early on in the project, I had brought my wok, a few pots and pans and of course my favorite kitchen knives to store in my little studio apartment. The lessons that I had learned from a few years of restaurant cooking stayed with me and I loved to be creative in the kitchen. It was a little bit of an escape from everything else on my mind at the time and after all, there's just something primal about slicing up raw meat with a sharp knife.

The gang would often hang out at whoever could get the larger studio apartment and there we would discuss either the injustices of the day or the exact amount of garlic to add to the Chinese Chicken. Carlos always insisted on more garlic. I sort of became the group's traveling chef and I often raised the eyebrows of other apartment folks as I walked through the complex with my pans, knives and spices.

Our gang managed to kill a few bottles of wine one night, following a work day so bad, I was doing voice impressions of our British project manager while still in the office. Maybe I intended for people to hear me, maybe not, but I had to be reminded that we were there after all to help this customer not make fun of them. Even the employees within ear shot thought it was pretty funny, but I digress.

Dinner would always receive polite compliments from our group no matter how glorious or disgusting it actually was, but after a mentally draining day they would have eaten just about anything. "Tonight we are serving Pork Hoof, boiled and served with rosemary and mint, in a congealed mutton sauce. The vegetable is a cubed parsnip infused with pureed brussell sprouts, and chopped oyster mashed potatoes with strawberry jam". They'd eat it. Bet money on it.

As we bonded, we used to find little mental diversions to take the focus off the day's problems. We would do things like define our own units of measure. You know, weights and measures used to define a situation. Like…If you could make it through dinner without being insulted it was a good evening. If you had to order the "hangover breakfast" the next day, it was a good night. 2 to 3 bottles of wine between four people at dinner meant it was a really bad day. Margaritas at Senior Frogs around 4:00 pm, meant it was a really, really bad day, but you still had to go back to work. Ashton cigars in the park at noon, meant we were thinking about not coming back to work. The brick oven pizza at the Rock Bottom for lunch meant we probably would be working late and missing dinner. Finally, cooking dinner for our own team at the Residence Inn, meant we needed a bitch session.

Hideaway

In the middle of downtown Milwaukee you'll find Jane and Darryl, two of the finest people you'll ever meet, and they own a pharmacy on Jefferson Avenue. They alone are the reason the business is successful, and they carry the term "personal service" to a whole new level. Even with national chain pharmacies all around them, they keep their customers by remembering their names, their prescriptions, and whatever knick-knacks and add-ons they like to buy. It is perhaps the most efficient use of retail square footage I've ever seen. And, you can get a really good cigar there too. I preferred the Ashton's myself.

The pharmacy was a safe haven for me when the day got to be too much to take. The job was always stressful especially when you're an external consultant and have little control over the corporate decision making, but that is the nature of our business. Darryl and Jane always welcomed me into the store, and it was a wonderful escape to stop in and grab a diet Cola and a chocolate bar. Yes I'm aware of the oxymoron associated with my choice of snacks, just let it go.

The Train

One night on the way to a dinner evening out that Jane had planned, we made a stop at an eclectic little place called simply "The Train". I can't remember the exact location, but "The Train" really is what it claims to be, a well maintained little lounge car from a now defunct rail line. I say lounge, because when we first walked onto this rail car, an older gentleman was standing near the entrance with a white jacket and bow tie. We stopped and I asked him where the bar was.

His delayed response was "We have no bar here" <long pause> "We do however have a lounge" and he extended his left hand to a narrow passage leading to the bar, er, I mean lounge.

I place the motif at circa 1940, but I wasn't around then so who knows. The art-deco table lamps were dim, the seats were a deep red crushed velvet, the wood paneling was real and supported occasional chrome push buttons behind the seats. The push buttons were labeled "SERVICE", but of course we didn't know what that meant. For all I knew this could have been a brothel, and the "SERVICE" button meant you were about to receive just that, SERVICE.

After ten minutes of waiting, I decided to go back to the entrance and find our "host" who was in fact stationed in a little cubby hole of a liquor storage area. I started to give him our order when he quickly interrupted me.

"Have you pressed the "SERVICE" button sir?" OK, I get it now, role playing. Who would have guessed that these 60 year old buttons actually worked. In the immortal words of our dear Jane, "What.. ever". We pressed the "SERVICE" button.

The porter was about 70 with a reddish brown hairpiece from the Johnson administration, wearing formal attire for a waiter even in those days. He carried a white linen napkin draped across his arm and a serving platter that carried a notepad and a gold pen. His look and demeanor was reminiscent of a black and white television image of John Astin playing Gomez Adams.

Same face, same hunched over stance, same mustache. He took Jane's order, he took Darryl's order, then he turned his hatred towards me. I was the one who asked for the "bar", I was the one who didn't press the "SERVICE" button…I was the one who wasn't playing the game…

"May I take your order <pause> Sir..".

I first asked for a Molson or Labatts Canadian beer.

"We don't carry that brand <pause> sir".

"Perhaps you'd care for a Brand…W or X or Y".

"No" I replied, but I next asked for a "Bailey's on the rocks".

His response was, "We don't carry that brand <pause> sir" but then offered "We carry Emmet's instead, We find it to be a much better product…".

I mean, How do you respond to that? "That would be fine" I said, and he nodded once to indicate that he finally had me where he wanted me.

Looking back, I didn't care much for "The Train" or the role playing game, but it was truly a unique experience. Dressing like you're a movie star from the 40's with a really bad toupee guarantees a pleasurable experience.

The Evil Breakfast

It was difficult to remember who actually coined the phrase "The Evil Breakfast", but it was a very appropriate description for a remedy to a common ailment. The reason I can't remember who may have said it first may have everything to do "why" we felt the need for an "Evil Breakfast" in the first place. It turned out to be the best cure for a hangover following a night of stress reducing alcohol consumption.

The Wells Street Café was a little dump of a place right around the corner from Jefferson and Wells. It was a six table joint with a counter, that also had a hamburger lunch special, although the only thing special about the hamburger was the price. Greetings and autographs adorned the back wall from semi-famous people most of whom had long since passed on. The tables and chairs were something you might have found from a garage sale 20 years earlier, and the dust-animals hanging from the ceiling ranged from bunnies to hump backed whales.

The entire place was a smoking section, and you couldn't leave without your clothes smelling like unfiltered Chesterfields. So why go there? Everything was fresh, and I don't mean "pulled out of a warming tray" fresh. I mean pulled off the grill, put on a plate and placed in front of you fresh.

The eggs were done perfectly, the hash browns were real potatoes grilled in tons of butter, and the sausage links were pulled out of the refrigerator and cooked right then and there. There was none of this "sit under the warming light and hope for the waitress" delay. Remember I said there were only six tables, so when the food was done cooking you got it, no delay. Amy and Karen would usually accompany me in the early morning hours before the work actually started, and it became a ritual after a late work night.

Nothing Evil about it, you might say? I guess you had to be there, to see the presentation, to see the plate. You had to watch how they greased the egg pan, the full ladle of melted butter-like substance they used to make the hash browns, and the spatula they used to butter the toast. It was a grease fest. One day we could actually take our knives and scrape excess gobs of butter-like substance off the toast and make a substantial pile to the side. Yum.

Typically still exhausted, we ate in silence sometimes, still trying to recover from the previous long work day and the evening's beer. The Evil Breakfast seemed to be just the ticket, just enough cholesterol, just enough carbohydrates, and complete with two large Cokes, just enough caffeine to start the day.

After the carnage was complete, you could feel your own blood pressure rise, the warmth rise up from your chest, neck and through to your head. It was the caffeine, we agreed than offset the effects of the cholesterol as the body seemed to reach a state of equilibrium sometime during the meal. With 2 aspirin for dessert, we were now set for the day. As visually horrible as the whole experience could be, the meal provided enough calories to last long into the day.

Yes it was evil, but it was a good evil, if there is such a thing. I found out three years later that they closed the Café in favor of more table space at the neighboring Sports Bar. The same square footage formerly dedicated to the cure of hangover now caters to the creation of the ailment. I shared the news later with both Amy and Karen and it was a sad day for all.

Greasy Ribs and Beer

Our work group had decided to grab dinner another night at the Brew City BBQ, only a short drive away from the office. Considering the fact that most of us were already tired from the day, and the fact that the Brew City BBQ doesn't serve anything that isn't messy, this may have been a poor restaurant choice. We started with beer which was a mistake, but then again it might be fun to see a few drunks try to eat a full slab with extra sauce. As expected it was quite a show, and the extra napkins the waitress brought in sympathy didn't reduce the "stainage" much. Sometimes you just have to laugh. It was fun to remember back to the days when the guy on the other side of the table with the reddish-brown stains on the white Polo dress shirt, was me.

We had accumulated a bunch of people from work around our table, a 50-50 mix of male to female. Just after dinner, but before the plates were cleared, we were visited by some friend of a friend of one of the guys at our table. This guy was a tall 25 year old African American in very good physical shape except for a bit of an equilibrium problem.

It seems that our new friend had been at the bar for a while and was struggling a bit to keep his balance. In any case the story here doesn't center around the new addition to our party, rather it focuses on the reaction of the women at our table while this guy was there. Sadly, it was a saliva fest. This guy, we'll call him "Joe" to keep it simple, was really "cut" as one of our customer-friends Annette had noticed and mentioned to the other girls at the table. Not that I know from guys. I mean guys don't really know what women find attractive aside from six-pack abs, but I gathered from the female comments that none of them tried to hide, that Joe was a good looking guy.

"Joe" was wearing a tight fitting shirt, which showed off the six-pack quite nicely and the girls seemed to be focused on Joe's midsection while he spoke. I guess that this is not just a problem for men, the women had trouble looking Joe in the eye too. It was interesting to look around the table to see the females react to this boy-toy.

One of them stared at Joe with her face resting in the palms of her hands, totally mesmerized. Another one with long hair was twirling it endlessly while another was bearing a constant shy smile and biting her lower lip while he spoke. It was like the rest of the males at the table had disappeared. Joe's assets included a solid educational background, good paying job, and extra money too from some magazine modeling he did on the side. Joe was sure to mention all of this in the few minutes he spent at our table. Joe made me sick, but he worked the crowd really well. I made myself feel better by believing he was gay.

Our new friend eventually made it back to the group he came in with, and immediately after his departure we could hear deep breaths and sighs coming from the women at the table. Some of them had to adjust themselves in their chairs. Maybe the sighs were stemmed in disappointment because none of them actually got his phone number, I couldn't be sure. Thankfully though, that was the end of the evening.

Gunshots and Pork Shanks

There are two other memorable restaurant experiences worth mentioning separately from the group, and not necessarily because the food was outstanding, but rather for the ambiance. On another cool summer evening, my two favorite Milwaukee business owners (Jane and Darryl) tried to get us reservations at this one steak house that everyone raves about once they've been there. "Coerper's" is a tiny little place deeper than it is wide, as the business frontage to the street couldn't have been more than 40 feet in length. There was a small red neon sign with the name of the establishment hanging in the window, and at first glance looked like a cozy neighborhood bar from the 50's.

The neighborhood surrounding Coerpers was in a very old working class section of Milwaukee, and many of the local buildings were boarded up and vacant. The parking lot adjoined the restaurant, and sat directly across the street from a Baptist Church where a band was playing some gospel music from the church parking lot. The Coerpers parking lot itself was surrounded by a 10 foot cyclone fence topped with barbed wire and was guarded by a private, armed security guard. I'm not sure how I actually felt when we first arrived, but it wasn't a comfortable, or content feeling.

Now I'll admit, I do travel quit a bit and you get into the habit of "surveying the territory" as a means to assess risk. It's a natural reaction really, and you find yourself doing it just as you would in any strange town, in any country. In this case, some impressions were well founded as I heard distinct gunshots from a medium caliber handgun maybe a block away. OK, so you might ask where I came up with that observation, or that claim. Trust me here, being from Detroit and experiencing the sounds of typical 4th of July celebrations there, it was maybe a .38 caliber or a 9mm round. The pitch was too low for a .22 or a firecracker, and not low enough for a 10mm, a

.40 caliber or a .45. I can explain how I know that stuff in another story. We found ourselves quickly making our way into this fine establishment for, of all things, a little dead, red meat.

Stepping inside the safe haven of Coerpers placed us directly in front of a long bar with dark woods and high polished chrome. It was clear that the place was old, and well maintained but never updated. It looked like a long bar from the fifties with bottles back-lit on glass shelves and where the likes of Deano, Sammy and Frank could have been sipping a "Dewars baby" and laughing about their last gig. We stepped back in time, had a drink or two and then made our way to a large round table.

Coerpers was known for steak, but everything else was wonderful too. Our waitress was very professional and courteous and looking at the prices on the menu told me why. The salad and potato was fabulous but the Filet Mignon that I ordered was pretty darn good. To this day I am not sure what they meant by "Charcoal Grilled" as a means of preparation but it was absolutely fabulous. I figured if you take a chance by coming here, your last meal should be a good one don't you think? Pack your "nine" and go. "Bust a cap if you have to".

"The Shanks"

Just to add a little European flavor to the story, we'll have to speak of another fine establishment for dinner, and another field trip to the outskirts of Milwaukee proper. One of our team had been talking about some little German restaurant off in the distance, away from the trendy downtown area. A place where the plates are like serving platters and the portions were big enough for two.

As it happens, this little out of the way German place would fit nicely into what your expectations might be for a little out of the way German place. It was cozy and the chairs were big and comfortable and the tables were big and round with plenty of elbow room. Remember we're thinking Western European here, elbows on the table are OK, hands in the lap mean you're doing something you're not supposed to. The walls were adorned with stained glass and trimmed in dark Oak. The central room had high ceilings and yes, even Accordion music piped in from speakers hidden behind plastic greenery.

Despite the challenge from the people who had been here before, I was not interested in ordering the house specialty, which was of course "Pork Shanks". Remembering all that they said about the portions here, I chose to order another German specialty that I truly love, called "Roladen". Roladen is thinly sliced steak rolled up with minced onion and strips of bacon then baked slowly for about a week. In any case, served with brown gravy and mashed potatoes, Roladen is a little slice of decadent heaven for me and I was looking forward to it. Two others at the table got adventurous and ordered the pork shanks. "How could you" I asked. "With ease" they replied and then we waited.

After salads, and chicken dumpling soup, the main courses started to arrive at the table. First came the "Pork Shanks" and yes, by all accounts the portions were huge and looked like they alone, could feed a family of four. In addition to the pork shanks, the plates came with a side order of massive potato and onion dumplings you could cut and eat like a steak. The folks at the table almost started to applaud.

Everyone was in sheer awe of the presentation and the volume associated with each dish. Finally my long awaited "Roladen" came and was placed in front of me. There was silence at the table as all had turned to look at my plate. Instead of a serving platter sized plate, like everyone else had, I received a regular sized round dinner plate.

On it, somewhere in the middle, existed two "Roladen" with a few vegetables added to sort of fill up the presentation. Lets put it this way, before I even started to eat, I was thinking about where I could stop for a burger on the way back. It was good after all, but compared to the sheer volume of the pork shank dinner… forget about it.

Finding the "One Thing"

Based on recent criticism from these same friends and co-workers, although they would merely refer to it as "observation", I would repetitively order one specific menu item at each bar or restaurant we would frequent. "Really? I hadn't noticed". They joke that it shows a lack of adventure on my part, while I defend myself by claiming to simply choose the "best" thing from each menu. And to answer the adventure part of their argument, I like to visit many places for variety, rather than ordering a variety from one restaurant. I prefer the venue change in order to experience the variety. So there. (Picture me sticking my tongue out here).

What was it that made the pub or restaurant experience unique or enjoyable? In some cases it was a specific dish, in others it was the atmosphere, still it could have been a combination of things. While thinking about past dinners, and parties, and "stress reduction evenings", I started to experience an involuntary chronology of memorable events, all connected to Milwaukee.

I remember "Eagan's on the River" for salmon or just about any type of fish. Of course, ribs at the "Brew City BBQ" but be sure to do your drinking after eating. The Chicken Picatta at "Louise's" is especially good when drinking a bottle of Kendall Jackson with Darryl and Jane. Meanwhile Jane grills me for good gossip after the consumption of the wine.

The "Lakeside Inn Café" is a great place for lunch, they always have good soup there and the caramel apple desert is to die for. You must go to Kopp's for fast cheeseburgers and a bag of fries while driving a convertible.

Casual dining winner: The"Third Ward Café" for anything made with pasta, the marinara sauce is truly awesome, and two of my favorite "non PeopleSoft" road warriors, Jeff and Carlos, made good company each time we went.

The "Rock Bottom" was the place for brick oven pizza. I also remember the Rock Bottom having a chef in poor mental health, having once come up with a soup called "Pumpkin Raisin Bisque". Please.

The food at a local art-deco hotel was fairly lackluster, but I particularly remember the way our waiter "Wendell" treated me when I tried to order something off their own wine list. At the top of his lungs, he proclaims "Oh God, I wouldn't order that" for the entire restaurant to hear and of course respond to. I waved to the restaurant crowd in acknowledgement.

It wasn't the cheapest white on the menu by far, and it was after all, on their "freaking" wine list. I took "Wendell's" alternative wine suggestion and reveled in the fact that I was the one picking up the tab for the evening. Wendell would get what he deserved later in the "tips" section of my American Express receipt.

Chapter Eleven - Back and Forth
to Europe

Near the end of August 1997, I was contacted by a friend of mine managing the consulting operations for Benelux out of PeopleSoft's Amsterdam office. He had asked if I could take a break from our Milwaukee customer and head over to The Netherlands for a week to assist in a pre-sales effort for a large customer. They were apparently planning a multi user demo on PeopleSoft 7.0 but there weren't many technical people there with experience in the new architecture.

Europe was just starting to build consulting operations then and the PeopleSoft customer base was just starting to grow. Securing this customer would be a big win for the Benelux team. For the demo, there were some outstanding performance issues with the database, and they couldn't bring up the BEA's Tuxedeo functionality for "remote call" processing.

Another functional consultant and I got the green light to head over for a week, provided we kept on top of our respective teams back in Milwaukee. Based on the internal rules for international travel at the time, we were allowed to travel business class for any flight longer than 6 hours. That made the trip palatable after all, when the seat allows you almost a full recline. Oh, and the food is pretty damn lovely as well.

The comedy of errors began as soon as we arrive at Schiphol International Airport in Amsterdam. Our contact there told us to "hop a train to Den Haag" and he would pick us up at the airport. "OK, which train station in Den Haag?" "Where do we pick up the train at the airport?". It was all clearly defined. Yeah sure.

We called our local contact who began to speak to us as if we were familiar with the airport and all the train stations in Holland, but eventually we came to some sort or understanding on what to do and where to go. I think back on it now and it was all a blur, but we did find our way to the airport station and then on to the correct train station in Den Haag.

We arrived at the small station, but trying to confirm that was a little difficult. The way our contact pronounced the name of the station and the actual spelling on the plaque at the station platform were not an exact match. This all from an English perspective anyway. Proper names we saw on street signs and station plaques were filled with "hard" consonants and not too many vowels. I came to the swift conclusion that Dutch was not an easy language to learn, speak, read, or listen to.

The trains here were modern and clean but the rails were rough in spots, reminding me of New York or Chicago. The grab handles in each car were not there for décor. If you were standing, you needed them. The Den Haag train station had been there for decades but had a charm imbedded in the dark brick and old oak doors. Standing at the front of the station with bags safely in hand was the real first chance we had to breath since getting off the plane at Schiphol. Only then did it hit me exactly how far away from home were.

We found a local shop on the corner across from the station and attempted a phone call with international calling cards. Aahhh… not so fast you technology assuming American you. The public phone systems in some places here were privately owned and on smaller networks. This phone took money to make any call, even a toll free one. After a frantic search for a way to exchange some paper currency for coin, we were able to call on our local sales contact to get picked up.

Piet from the Benelux office was a good looking, dark haired typical sales guy and he briefed us on exactly where we were with the customer and the demo. It appeared that we would be starting that day, and trying to fix the technical problems the minute we walked in the door. Not a great situation to be in if you're planning a demo to a potential customer on the equipment that wasn't working.

We made it to the customer's very modern facility in the center of Den Haag and were introduced to the entire "soon to be named" project team. I met the technical lead who assisted with the install of the UNIX equipment and the database and we headed off to a workstation to see what was what.

Johannes was an older, tall and lanky man with thinning grey hair and a fairly good command of the English Language. He struggled a bit with certain words, but then looked at me from time to time to help complete his sentences for him. We started reviewing the system configuration together and found out a couple of simple mistakes they made following the instructions for installation. Apparently our installer left with the database working properly, but then the customer chose to move the system to another set of servers on their own which messed up the configuration.

We also found that the Process Scheduler (a background running program that looks for batch job requests) kept shutting down on its own. After we cleaned up the large file system that was overwhelmed with trace log files, we were back in business. All that was left to do was monitor the system while the rest of the group beat the crap out of the system during the demo.

We broke for lunch after 1 PM and headed to the company cafeteria. It was just me and about 10 other technical guys from the customer. I tried to make sense of the cafeteria stations and found the bread and individually wrapped slices of meat and cheese. I grabbed a bowl of clear broth soup and headed to the table with the rest of the gang. I was the taller one at the table which I think surprised most, even prompting one comment in broken English about my height. Most of the males in The Netherlands were over 6 foot tall.

I watched as some of them had selected a single piece of bread and a single slice of meat or liver sausage spread across it. They carefully cut it with a knife and fork into small squares and ate it a piece at a time. "Interesting" I thought as I piled my ham and salami and cheese and mustard into my croissant and picked it up with my hands. One of the younger men across from me started laughing and called me "Condiment man", apparently never seeing anyone make a traditional American sandwich.

"If you think this is funny" I said, "Go to America and order something from Subway".

The conversation loosened up and we started sharing stories about each other's culture and foods, entertainment and politics. It was friendly, and they were truly curious. I sensed that many of them, much like me, had never traveled outside of their own country.

The day ended without much incident and we were escorted by Piet to our hotel outside of Den Haag in a coastal suburb called "Schreveningen". We were booked at a nicer than normal hotel right on the North Sea and connected to a boardwalk that led to numerous shops and cafés to the right of the hotel. "Nice Digs" I thought as we checked in. Piet was staying there too to avoid the commute to Amsterdam during the week. I met Piet for dinner after we checked in.

My history lesson began with the first Orangeboom beer and how Holland (The Netherlands) has been driven to nationalism and almost into isolationism the past 100 years. Especially during the last world war, with France and Germany bordering their small country, Piet told me stories of how the Dutch used trickery before they were invaded to identify German spies at their Eastern border.

In one example, Piet talked about how difficult the Dutch language was to pronounce correctly and this was used as the basis for a test at the German crossing. Someone crossing at that border was simply asked to pronounce the name of the city we were currently in.

"Schreveningen?" I asked.

"Yes" Piet replied, "but that is what I mean". Piet proceeded to pronounce the name in Dutch which blended consonants and a unique sound that spoke like phlegm in the throat.

"Only a true Dutchman could say it correctly". With that, Piet told me that the ones who messed it up were captured and sometimes shot at the border.

After a while I began to understand why this country, after being invaded by Germans twice and French as well came to take such pride in their country, culture and language. Thanks Piet, for the history lesson. It became a trade. I tried to explain that not all Americans are idiots, and that many of our foreign policies are set by small groups of politicians only after they are elected. Then there isn't much we can do except wait for their term to be over so we can vote them out. I was surprised how much he and others in Europe knew about American politics and how we know so little about theirs. Ah well, cheeseburgers were on the menu and the beer was nice and cold so the evening had to be a success. No political scandal was born.

The rest of the week went without incident except for occasional tuning of the database and clean up of log files created as the demo crowd created reports and general ledger edits and post processes. Then the news late one evening described the death of Princess Diana in Paris.

I expect that she was more beloved in Europe than in the US, and it was clear the next day that the people in town and on the streets were talking about it. The newspaper headlines captured pictures and stories from France and the UK and the television news even though in Dutch, kept the story rolling all day. It was fitting that the North Sea was calm that day and the sky was overcast with light rain.

Amsterdam

There was an extra full day in our schedule before we had to fly out of Schipol Airport, so we wrapped up things in Den Haag and boarded a train back to Amsterdam. Walking from the platform into the lower floor of the train station, I was immediately struck by the strong smell of "pot" everywhere. I inhaled a bit more deeply and Piet looked at me and just laughed. "Its good to be home" he said.

From the main train station near the city center, Piet and I boarded a bus and a few minutes later, were standing in front of his apartment just off one of the main arteries leading to the city center. Piet's place was two flights of stairs up, taking the third floor of a traditional Dutch 4 story house which was converted into separate flats. He had a completely updated kitchen done in simple white tile and a small balcony facing a city park, filled with mature hardwood trees. The entire flat was decorated in white with except for the hardwood floors that must have been 100 years old but still maintained their warmth.

We unloaded our gear and made plans to head to the red-light district just after a visit to a local pub/restaurant around the corner. Piet explained that the locals typically stayed away from the red-light area, because it seemed to support sporadic waves of street crime depending on which drunk or drug user needed money. Still he figured that I might like to see it and of course the "Banana Bar", that I had mentioned I heard about from a friend who had come here years before.

Piet received a cell phone call from another local PeopleSoft consultant that I happened to know and we made plans to find each other at a corner bar called the "Salty Dog" on the canal near the "Banana Bar". After dinner but before heading to the red light district, we made a stop at what looked like another bar and Piet led us in towards the bar in the center. The place was called the "Bulldog" and Piet asked if I happened to be a fan of Van Halen.

"You might know what kind of place this if you remember the lyrics from their song 'Amsterdam'." Then the smell hit me. Piet told me to order a coffee and ask for a menu.

"The pot menu".

"Ah yes. Columbian Gold, Panama Red, Hash". It was all here.

Since my high school days were long since over, all I really wanted from the place was one of the cardboard coasters they provided with Piet's coffee, so I pocketed a clean one from the ledge at the back of the bar top. I would later have trouble explaining to Cheryl what it was, why it was significant, and how I avoided ordering something else from the menu.

"No honey, I stayed relatively clean on this trip".

"Relatively?" She asked. "What the hell does that mean"

Although I was only referring to alcohol, she had trouble dropping the subject.

The Salty Dog was a dive bar on some corner of the red light district, but we had to walk down one alley poorly lit by black lights to get there. You see, white clothing, more specifically white underwear appears to glow as you feature it in a 5 foot glass window facing the alley or street. I've heard about this but expected red lights, not black. Some windows were empty, others were animated as the glowing young ladies struck poses for the benefit of those passing by.

From our bar stools at the Salty Dog, we faced a corner by the alley and the street in front of the canal. We enjoyed another cold beer and watched people as they walked or stumbled by. "This beats Vegas for people watching" I said to Piet. Soon our mutual friend from England, William had joined us.

I knew William from a previous customer and it was good to see him again. On the corner appeared a young black woman who was talking and smiling widely at a young man in an oversized and crumpled brown sport coat and jeans. He simply shook his head and walked on when another young man rounded the corner close to the wall and almost bumped into her.

She reached down with her right hand and grabbed his package in such a way as to make him nearly jump backwards. Whatever she said made him smile but he shook his head as well and walked on.

The three of us agreed that she probably couldn't afford her own window and this must be her way of "handing" out a business card. Sort of a "hi, how you doin" welcome to the red light district. We turned back towards the bar as Piet went for another round, and she was gone. Apparently she had successfully "handed" out her business card to a prospective buyer.

After a couple of tall ones, we strolled the walkway and there it was in all of its glory. The "BANANENBAR". I had heard stories of what took place in this fine artistic establishment, and I was told that "I had to go there". Of course this was by a depraved former boss. Piet and William looked at me as if I were the typical tourist that I actually was, and said that they had a better place in mind.

We wandered a little further and strolled into a "Live Sex Show" club, or at least that was the billing on the poster outside. It was dark and the carpeting felt a little too wet for me, like it had just been steam cleaned. I simply didn't want to know.

We sat in the back of the bar away from the stage because all the seats in front had been taken by elderly couples. Perhaps this was the most interesting thing of all, "Old people" I had blurted out to William. In his pleasing British accent he explained "Yes, funny isn't it, it seems to be the norm here most nights. I don't know why but they're always here."

The pre-show must have started because I heard whooping and clapping noises coming from the crowd in front. A young rotund woman in a yellow ruffled bikini and Carmen Miranda hat appeared from our left and jumped up on the bar. That was a feat in itself given her weight.

She produced a banana from her fruit-basket hat and the crowd began to whoop it up again. A young lad at the bar was chosen, probably because he looked innocent, to help her with a few banana tricks she had in mind for her new attentive audience. I can't and won't go into any more detail here, but it was nasty. Funny in a bathroom humor kind of way, but yes, feel free to use your imagination on what she did with the banana and how she chose to involve the young man at the bar. Simply nasty.

After the "warm up" act, the male and female leads of the show came bounding out of the door to the left of the stage, to begin all sorts of gymnastic moves… in the nude. There were certain things they could do that most human bodies could not possibly attempt without a trip to the chiropractor.

"OK, saw it, been there, done that, cross that off the life list". More shock value to cap off the night in a strange part of a centuries old city.

As we walked back to Piet's apartment he reminded me that it was mostly the tourists that went to that part of town. He then took William and I to a very nice pub with a pop music cover band playing in the back. The chairs were comfortable and the beer was served in unique glasses depending on the brand chosen. The end of the evening was comfortable, but the memories of our walking tour stuck with us. The week finally ended and it was time to get back to Milwaukee for the remainder of our upgrade planning to release 7.5. There were still other development efforts underway in Canada and we would be supporting them soon in production.

The months continued and our upgrade issues were resolved. We found the project team in Canada was great to work with and the initial go-live went relatively smoothly. There were a few performance issues related to network bandwidth, but they were identified and eliminated after they connected a dedicated T1 line between the Milwaukee and Toronto data centers. Work continued through the end of 1997, as we started to get advance materials for the new product release on PeopleTools version 7.5. The architecture now called for a strong middleware engine for transaction processing and added a layer of servers to our architectural solution. It made things more complex, but the Tuxedo middleware engine allowed for decreased traffic over the wide area network. It would allow a centrally located data center to support Europe on a UNIX/NT platform. Back then, it was exciting stuff.

Continuing the Upgrade Project

We had been working extremely hard to finish the modifications to the system and resolve the remaining issues encountered with the product upgrade. Before the system testing began, the management team decided that a small group needed to venture out to Brussels to meet with the European management team and make sure obstacles for the go-live had been removed.

The timing was also good as it coincided with the PeopleSoft European User Conference in Paris, 1998. It was a brilliant opportunity to hear what other global customers were doing with their solutions and cover our Belgium management meetings as well. One trip over the Atlantic always beats two.

I had been working closely with Matt Reinhart, the customer's technology manager who was about my age, married with two small children. We had similar winter sport interests and similar family backgrounds and hit it off immediately. He had a dry sense of humor, and fit well with my own. Sometimes you could cut the sarcasm with a knife when we both got started, but somehow the work still got done. Our mission was to evaluate if the European data center and technical team were ready for the new functionality about to be delivered over the network.

We knew it was going to be a long flight into the Brussels airport but really didn't know how difficult it would be sitting back in coach. We had several conversations and most were about business but some conversation was about family. It was a great way to get to know each other given the fact that we both couldn't sleep.

We both had a chuckle or two over the current project manager who was a short little man with thick grey hair, a type "A" personality and a strong British accent. Not quite cockney and not quite aristocratic, the project manager's voice was easy to imitate or at least make fun of. This was a guy who could summarize all our complex and varied technical issues with a statement like "Right... lets just get on with it then... ay?". We suspected he spent some time in Canada as well.

It was a good team of technology people, all with different backgrounds but dedicated to their company and fun to work with. It was true with most people here. You could head to your favorite gathering place with a group from work and get offered a beer from someone you barely know before the night was through. It was my kind of town. I picture Frank (if he was alive) changing the words to the song..... "My kind of town.... Milwaukee is...".

Matt talked about his family and we shared similar stories about working too much and never getting home to enjoy the people who mattered most. His was a different problem. Although he had a "regular" local job, he worked extended hours all the time which kept him from ever getting home for dinner or seeing the kids off to bed for that matter. He might as well be working for us, I thought because at least you had the whole weekend to spend with them. Working in Milwaukee, Matt was on-call over the weekend too.

The sun eventually rose on our DC-10 as we crossed the Atlantic and started our decent into Brussels Airport. Matt and I managed to stay awake during most of the night flight and felt pretty good at the time. We would pay for our avoidance of sleep later in the day as meetings were scheduled from about 10AM local time through 6 PM.

It was all I could do to keep my eyes open after lunch in Brussels. The meetings dragged on and although everything was done in English for the sake of the Germans and French and of course us. The only trouble was that it was the European managers who chose to use the meetings as a forum for airing out their dirty laundry. It was all very civil, but not very productive.

That night Matt and I ventured out from the Hilton Hotel near the Brussels center plaza or "Platz" where all commerce and trade used to take place centuries before. We walked around the square and eventually found a street lined with Cafés. It was hard to read some of the menus on the street because the languages changed from French to Flemish, but we settled on a small place where our table faced the street.

We were in no rush for dinner which was good, because restaurant service here took on a whole new meaning. I had read about this before though not experienced it in the UK. The people I met in the UK ate fast because I think it interfered with the limited time they might have to consume a pint or three. The rest of Western Europe, as the Travel Channel informed me, enjoyed the evening meal as a social event. Meet with family and friends, enjoy their company, drink and eat slowly, take breaks between the salad and main course.... I was prepared for my Café experience on this narrow little cobblestone street.

Somehow we made sense of the menu, picking out key words we could interpret and ordering the best we could in front of our extremely patient waiter. "Potage" for soup pronounced "potaagsh", and "Viande" for steak. Soup was supposed to be the first course, but we had ordered a bottle of wine to start. A different waiter brought us each a large flat white bowl and placed it in front of us, then followed by a silver plated but empty soup tourine.

Matt made the comment "I guess we get the soup first".

Just then our primary waiter hovered over our table, quickly and brusquely grabbed the tourine and bowls and shot a dirty look at the other waiter who placed them there. The waiter stormed off and Matt and I shared a quizzical look. Matt finally offered a quote that had me rolling off my chair.

In a calm manner, Matt had placed his elbows on the table, folded his hands together, leaned forward and said quietly "That must have been the Potage Invisibaal".

"Good too" I offered. "Subtle flavors".

"Yes, very light" Matt concluded.

The rest of the meeting schedules for the week went better than the first, and we reached agreement on several key points needed for us to move forward with the project. The days were long, mostly to take advantage of our time while in Belgium and we were exhausted. We went out in a larger group the next night and saw a few obligatory tourist attractions like the "Pissing Boy", but the big event of our trip was about unfold the following morning.

Our little band of travelers had booked tickets on the Thalys speed train from Brussels to Paris and would be there in all of an hour and 25 minutes. The European User Conference that year was booked at perhaps the most glamorous of destinations, showcasing the best in European culture and architecture. Disney World.

On the train we joked about the conference location and made fun of our own culture as the French experience "America" by visiting an amusement park hosted by an overgrown, talking mouse. During the conversation someone noticed the view outside the window as the train started to pick up more speed outside of the Brussels city limits. It became more than a blur. In fact you could feel your eye sockets start to ache while trying to follow points on the landscape.

Everything was moving by too fast to follow. On a plane the same dynamic didn't apply because while in the air, all other points of reference are too far away to move by your window quickly. Eventually, the train submerged below grade and the blur of the dirt and grass on either side took on a quality of wallpaper. We all went back to planning our upcoming meetings and possible evenings out.

Paris (Sort of)

It was just like Disney in Florida. Beach style hotels with nautical themes and female workers running around in women's dresses from the 20's and 30's while the men wore horizontal striped sailor shirts and white sailing caps. You could actually rent Swan paddle boats for the man made-lake in the center of the complex.

We checked in and planned for our on-site meetings and sessions at the conference. Others did the smart thing and located the on-site train station that led to Paris. The first day was all business, but Matt and I would wind up the long day of meetings with a couple of beers at a "pub" on the Disney property down the boardwalk from the hotel.

The next night we opted for a more authentic type of French cuisine that could only be found apparently at Planet Hollywood. Our entire group of PeopleSoft and customer folks gathered into one group of 14 and entered the Disney World theme park in search of the "House of Schwarzenegger".

Along the way we did a couple of theme rides, and found a way to grab some more cash from a conveniently located American Express booth. All along the midway, there were hamburger joints, pizza parlors, and other fast food choices that I am assuming Disney wanted the Europeans to think was representative of American cuisine. "Ah well".

In a way, "The Planet" was a bit refreshing in that you at least knew what to expect when you ordered from the menu. It was similar to the ones back in the US, except for the price, and the food was better than average. All in all our loud and a bit unruly crowd had a good time at "Arnold's" place and we wrapped up our European conference with good information to take back home, and a couple of memories to keep.

We had all arranged to stay two extra days in Paris after the conference concluded, for any follow up meetings with our European team and of course to enjoy the perks of business travel to Europe. The train station our gang found earlier in the week was located again as we all struggled to tote our luggage from the hotel to the Disney Station. The train ride to downtown was reasonable, and we wound up talking about the project and the Western European implications most of the way there.

From there the Paris Hilton was our next destination and I don't mean the cinematic blonde teenager. It seems odd to me now that she may have been named for a hotel, but then again with that kind of money I guess I'd be happy to call my son "Palmer House" or "Waldorf".

The actual Paris Hilton Hotel was in walking distance to the Eiffel tower and just about everything you might want to see on a short trip to this city, so the location was key. We met as a team for the last time going over the information we had gathered at the conference and assigned documentation duties to various members. I summarized the technical information and some risks we would expect to see with large on-line transactions over the wide area network. Everyone doubled up with their own sub-teams and we finished as much as we could before the end of the day.

Our group split again into smaller groups mostly because when trying to organize 14 people, its as difficult as "herding butterflies". Once you catch them, its impossible to get agreement on where to go for dinner, or what to wear, or anything else for that matter. Matt and I and two other guys decided to walk along side the River Seine and check out the lights and the sights before we all dropped from exhaustion.

We eventually found a sizeable yacht docked alongside the river that was actually a floating restaurant. It looked like an older motorized yacht possibly from the 50's with plenty of darkly varnished trim and planked deck boards. There were two levels of tables, but we chose to sit on the top deck overlooking the river. Matt and I both ordered stuffed Trout which was perhaps the best I've ever had, while the other two struggled to find something "safe" and chose a pasta dish. White wine flowed freely but nobody really went too far, we still had to walk back after all.

The next day was a free day for everyone and again the team split up into their own groups. There was the museum crowd, there was the tour bus crowd, and the remaining few of us opted for a walking tour of the Seine, the Champs-Elysees and the Arch de Triumph.

We started by hitting the Eiffel first because the hotel concierge suggested it before too late in the day "if we wanted to take the elevator up" he said. The last steps could only be taken on foot and the climb only seems easy from the ground. "It can't be that tough" I remember blurting. Now try steadying the camera when you're breathing hard after the climb. The view was foggy that morning but still impressive.

On to the river walk we managed to stumble across the "Musee D'Orsay" and noticed the poster on the building featuring a Monet and Manet exhibit. "It's a limited time thing" I said, and the rest of our mostly tech-geek crowd agreed to go in. It was an amazing comparison of styles and backgrounds and a once in a lifetime chance to see this volume of work in a single place.

From there we walked on, stopping for water or snacks on the street wherever we needed it. Eventually we found our way down the Champs-Elysees and did a little window shopping. Again, we thought we needed the "Paris Hilton" kind of money to shop there but it was nice to stroll down the same path that played host to so much history.

Again we tried to herd our butterflies for a later dinner, but everyone seemed to want to go to a different place. Half of the group split off and the rest of us opted for a café near the Eiffel Tower. The chilled Beaujolais was wonderful and the food was again a thrill. The chicken dish I ordered was cooked slowly in a small Dutch Oven and served in the clay pot with a subtle mustard sauce at the table. Yummy. After dinner we tried to snap some photos in the dark as the Eiffel Tower was lit by a digital sign counting down the number of days until the year 2000.

The following morning was our last in Paris. I found another street side café with Matt and I both had one of the best ham and cheese omelletes I've ever had. I don't even want to know how much butter was in it. We all gathered at the hotel a little later and piled into cabs for the trip to Charles De Gaulle airport.

Little things stick out in my memory about that short trip. At small cafés everywhere, there were chilled ceramic cruets on the table for whatever house Beaujolais they chose to offer. It was wonderful. The fresh tomatoes provided in a "Salad Mixte" were the best I've ever had, and the people here were mostly friendly if you took the time to pronounce a few words of French.

The trick I think was to "try" to appreciate their culture and language. Saying "merci beaucoup" seemed to go a long way there and aside from a few rude people, the trip was enjoyable. Current politics aside, it is a beautiful country to visit.

Coming home was more of a challenge. I would guess that most spouses that had never been to France, were probably not too happy that their other half got to go without them. We were asked by the customer to go as part of the project effort, but I can also understand why my wife might not want to listen to me describe how great the trip was.

The trip to Europe was getting paid for by the customer, and Cheryl got to chase three pre-teens around the house for a week without my help. If I have to explain the sensitivity of that situation any further, you are not married with children. She was good about it in the end and I would ultimately pay her back years later with a long trip to a country of her choosing.

The End of the Milwaukee Road

As projects went, this was a good one. For the most part we had stuck to an implementation schedule and rolled out more global supply chain functionality to Canada, the UK, Belgium and The Netherlands than anyone else had done by 1998. We made friends with our customer, we didn't just work for them. We met their families, we had dinner out, we traveled the globe, we rode snowmobiles together.

It was sad to leave this customer, these people and this town. One of our own sales people who was new to the PeopleSoft, was guilty of making some wild promises to the customer about some new product and its performance. He unwittingly forced me into an adversarial position when he claimed that we were not doing all we could to tune the system the way it should be tuned. It was a way for him to make it look like he was in command of the situation, but it alienated me and my team who were actually tuning the system for the development team in Pleasanton.

I chose to leave rather than damage his new sales opportunity with the customer. We had a wild party at the end of my last week. The project director and the entire management team showed up bearing fabulous gifts and embarrassing praise for the work that I had done gladly. It was a privilege to be there almost two years and extremely hard to leave. I will remember them all, especially Matt.

I miss Milwaukee in the Summer and the festivals on the waterfront. I miss the little restaurants, and the micro brews. I miss skating the 1000 meter oval at the Petit Center. I miss the Brewers games and the goofy little guy sliding down into the beer mug. I miss the smell of yeast filling I-94 as you drove past the Sprecher brewery. I miss the dinners we cooked up at the last minute. I miss the hot dog stand outside of NML on the way to the waterfront. Most of all, I miss the people.

Chapter Twelve - Dot Com Wealth
and Other Great Lies

With PeopleSoft 7.5 came a viability of options for some of our largest customers to globalize their software solutions. This meant bigger and more complex technical environments, more servers, more software for managing the servers and more production databases. The boom was on in 1998 and it seemed that everybody who needed to upgrade their current software systems, or everybody who wanted to dump their old systems was on a fast track to do it before the feared "Y2K" date. There was as much business as anyone could want.

During this time frame one of the largest segments of our business was found in the education and government space. Many state governments found our HR software to be flexible enough to meet their needs and even divisions of the federal government signed on to use PeopleSoft as well. The commercial sector was lively too, and we were running from customer to customer trying to keep everyone on track so they could meet their deadlines.

There were opportunities on the horizon for the technical people outside of PeopleSoft. Many of our best technical people both in the Pleasanton offices and out on the field left for potential riches during the dot-com boom. Some of them made money quickly by cashing in on options right after an IPO, others didn't do as well, sticking to profit sharing plans that existed one year but faded the next.

It was a huge draw when you heard the stories of instant wealth, but then again you had to be in the right spot at the right time and you needed the resources to support whatever business plan you were implementing.

I knew of one such company well known on the west coast, who bet heavily on moving their mostly mail-order business to the web for order processing. The hype was huge, and the advertising was extensive to get the consumer to move to the web for all their specific mail-order needs. But… I love the "but" in a story like this. The technical architecture had not been built and tested for the volume of transactions they might receive. Furthermore, there was no connectivity from the web storefront to the back office operations. In laymen terms, even if you could get into the website to place an order, it wasn't automated so that the product could actually be shipped. The whole process backlogged, orders were lost and the company almost went out of business.

Hmmm.. poor planning perhaps? Poor architecture? A little too aggressive to jump into web solutions?

Some of our customers were looking for ways to take advantage of the web as well, but much of our software was back-office in nature anyway. Nothing too "customer facing" at the time. General Ledger, Accounts Payable, HR, and Payroll, all background company support functions. Our product strategy people were planning more customer and supplier facing solutions but those were being planned for PeopleSoft 8, the next major release.

Consulting rolled on for me and many of my assignments were shorter term with a limited amount of time at the customer site. It seemed to be a function of helping a customer get started with their project, planning technical architecture and moving on to the next customer. The side effect of this was that I was home more often.

I set up camp in our Southfield office and even though it was an hour drive to get there, it was nice to have access to our network, a clean bathroom and a way to get home and sleep in my own bed. Cheryl liked it too and made it a point to mention often, that she liked me there. I would help on proposal work and business development activities from the office, helping out the regional management wherever I could. Then it would be off to another customer site for a two or three or four week visit.

I worked with a couple of other managers in defining a process for evaluating a customer's technical architecture and making recommendations for the support of new PeopleSoft product or an upgraded solution. We put it together as a packaged service and trained a few people on how to deliver it. It was a huge success.

Easy for our sales guys to sell, and a load of value for customers, answering a lot of questions about servers and software in a short amount of time. This became the new vehicle for helping a customer get started with technology when they started a new project.

I also worked with another small team of people who were building a proprietary implementation methodology for PeopleSoft called Compass. This was a series of activities and steps that guided a customer through an entire implementation or upgrade. It helped create a full project plan and even helped to indicate the type of resources you would need for each activity. Since my background was architecture, I added content to the methodology for a number of missing areas like capacity planning, disaster recovery, help desk, software change management, performance risk mitigation and other activities.

It was great to work on things from the local office. I had forgotten how big my kids were getting. My youngest daughter was now 5, and I had been mostly away from home for he last 3 years of her life. She seemed distant, more so than the older two but as I spent more time at home she provided more hugs and conversation. As much conversation as you can have with a seven year old that is. Our oldest boy was almost 9 and the middle daughter was 7. We planned as many dinners out as we could and did as much as we could on weekends. The time home during the week helped strengthen the family and I was grateful for that. It wasn't just Cheryl playing the "enforcer" on homework, or picking up their clothes. I could take some of the heat off her and be the tough parent too. "Fun-buster Mom" she would call herself.

Our stock price was doing well through 1998, but sliding along with the rest of the technology sector in 1999. Everybody in just about every technology company panicked and was looking for things they could do to stop the slide. There was nasty commentary coming from wall street directed at CEO's and CFO's all through the Silicon valley. Whether that was a causal factor or not, the board of directors had decided to make a change at the operational top of our organization and the whole of the company was shocked.

Craig Conway was hired and groomed to become the CEO of PeopleSoft in May of 1999. In a single move we went from jeans and t-shirts in the office with Dave Duffield to suits and ties in the office with Craig. It was a major culture change for the employees and a number of key players left the company quickly with sadness dripping from their good-bye e-mails. Dave remained on the board of course, and stood by Craig as he made his initial appearances in company meetings and user conferences. There was a conscious effort by the new management team to appear more professional, more of what Wall Street expected from a billion dollar company.

Slowly perks and benefits started to change. The Monday/Wednesday/Friday bagel deliveries stopped, preceded by a memo from Craig describing the annual cost of such a "perk". Health care and other benefit plans started to cost more across the board whether you were single or married with children. Carry over vacation days were limited. Local office summer and winter budgets used for picnics and holiday gatherings were cut. It was the little things, in fact it was the bagels that seemed to piss most people off.

We were all good corporate officers and understood that this is what you do to control costs. We wanted to see the stock do better because most of us were stock and option holders. If we could save a penny a share with expense savings... so much the better. We may not have been fond of Craig and the recent changes, but we still loved the company.

1999

The annual User Conference was coming around again in the late summer and I was asked to return for my second visit with two auditorium sessions that year. Every visit to the User Conference was a test for me and a chance to beat back a fear of public speaking. I would recall an incident in high school where I was asked by a priest to participate in a Graduation mass for the senior class and all of their parents. Skipping the painful details, lets just say that I froze, motionless and speechless on a stage in front of about 400 people. Thankfully I never had to see them again. Since then I had forced myself to feel more comfortable in front of people. I took sales jobs in college and volunteered for presentation work in my early jobs after college. Now with the PeopleSoft User Conferences, it was a chance to get back on a real stage with lights, and a live audience.

That particular year, I actually got a standing ovation or at least a nice round of applause while they stood up to head for the exit. I prefer to think of it as a magnificent performance. "Where's my agent".

After the conference was over, I was called out to Manhattan to lead the technical effort on the Beta install of a new analytical reporting product called EPM. Before the official release of PeopleSoft 8 had come out, we bundled some of the early Tools features into a new analytical reporting product and made it available to a few customers interested on working with it pre-GA, or before it was released (General Availability). This particular company was located in Mid Town and the hotel accommodations were a short walk away at the Hotel Intercontinental on 48th and Lexington.

It was my first time in New York and though I had experienced the concept of "fast paced" in other parts of the country and in Europe, Manhattan changes the definition. The people are unique there, because the whole city is unique. Not bad really, but there were two distinct impressions I took from meeting the people. Outwardly cold but inwardly warm. A visitor might initially walk away with the wrong impression of the city, because it can be cruel and impersonal on the streets.

Maybe my whole problem stems from the fact that I'm an outsider, a Midwesterner and not a native New Yorker. Maybe I couldn't initially love it there, because I'm not from there. The folks who reside in that part of the state are militant pro-activists when it comes to the benefits of living in or around New York.

Much like talking to folks from sunny California, The New Yorkers can't understand why anyone would want to live anywhere else. In conversations with several local folks about my first impression, they discounted my opinion saying that I wasn't "tough enough" to live here. Now I grew up in Detroit and I mean the emigrant working class neighborhoods of Detroit, and I've also worked and lived in a few other places around the world. This discussion is a matter of preference and objectivity, not toughness.

Back to the California-ites. Have you ever noticed that they refer to the hills surrounding the bay area cities as the "Golden Hills of California?" Could it be that aside from two months of winter rain where the grass is green, its all just dead grass? I rest my case on the objectivity argument.

I figure that everyone needs to form their own opinion. Mine started to form as soon as I had my first, second, third, fourth and so on encounters with New York cab drivers. Right off the plane at Laguardia I grabbed a cab, tossed my bags in the rear seat where a gentleman that needed a bath turned to me and said "Ya?". With that heart felt welcome, I responded with a specific direction to take the Midtown Tunnel and an address on 42nd street in Manhattan. Now I ask you, was that clear enough?

For my enjoyment there was reading material pasted up on the front seat back and therein was listed "The Taxi Passenger's Bill of Rights". I was comforted to know that "I had a right to a clean comfortable cab", "I had a right to a driver who speaks English", "I had a right to give directions to my destination"… the fictitious list of rights went on.

As we headed towards what appeared to be a bridge rather than a tunnel, and yes I travel enough to know what a bridge looks like,

I asked "What happened to the Midtown Tunnel?" He suddenly spoke no English. We took the bridge, and a tour of Manhattan eventually winding up in midtown after we added another $7 to the typical fare where I paid him the fare plus a single dollar. Suddenly he spoke English very well, expecting a tip in proportion to the fare. Suddenly I spoke French, and muttered something about a tunnel.

Hotel rooms in New York are very "European" for the most part, nicely maintained but small and expensive. Like all major cities around the world the prices rise in relation to the season and regional demand. I had been staying at one of Midtown's nicer hotels where the rates began to rise above the $300 mark after Thanksgiving. This was typical of the hotels at the time. After getting that expense bill, my customer decided to move me into an efficiency apartment. "Efficiency" in Manhattan means a room size akin to that of a Red Roof Inn, but with a stove. Ok, It was nice and I didn't have to share a bathroom with anyone else on the floor.

After the apartment had to be given up, my customer suggested that they be allowed to find a hotel for me. Maybe they were still in shock about hotel prices in their own town. I mean, when you live here, how would you know? In any case, they put me up in a nearby hotel on 42nd near Third. I could write an entire book about my first night's stay but I'll stick to the highlights.

At first glance it looked like a fairly modern building facing the street, mostly darkened glass, rising up 50 floors or so. You have to remember however that most of the buildings on this street were built pre-WW2, so the term "modern" is relative. In any case I walked past the beat up asphalt driveway that circled the entrance of the hotel into the lobby where I waited to check in and get my key.

I found it interesting that my room number was #2001. "A Space Odyssey" I thought, an easy number to remember when you travel every week and forget things like room numbers, rental car colors, and packing clean underwear. I entered the elevator to stand on what I'm sure was once plush carpet from the 1970's and noticed the subtle dark brown and red oxide colors reminiscent of "Leisure Suits" I wore 25 years ago.

The hallway carpet and wallpaper leading to my room, next to the ice machine, was also circa 1975 and showed plenty of wear and tear but no big deal. Looking at the door, I was caught by the fact that there wasn't a door knob on the outside, No handle of any kind. Once again I support my claim of shock being that I do stay in hotels every week of the year. I inserted my magic card, heard something buzzing inside this massive stainless steel rectangular box but nothing happened. I tried it again, saw a green light, heard the buzzing again but this time pushed on the door. "Ah Ha" I'm in!

Maybe I should have never turned on the lights. Maybe it was better to stay out, spend the night on the town, get hammered then come back to crash. It would have eased me into this retro look. I reminded myself that retro was back in fashion. What's the rule? The fashion is to go back 25 years? Whatever. Retro in this case also meant original. Original bed, original chairs, original color scheme.

The carpet had been replaced at least once because it was beige instead of brown. At least it was still beige in certain places like underneath the dresser. The entire carpet was covered with various and sundry stains, spills, and bits of color none of which I chose to speculate on as to their origin. I mean I actually was apprehensive about taking off my socks and walking barefoot around this room.

The chairs were old and worn, faded and stained. There were rips in the fabric on the arms of both chairs and clumps of stitching were coming out of the seams. The dressers and night stand were also circa 1970 with faux antique white finishes, chipped corners, and beige faux marble tops. The lamps, bedspread and drapes were patterned after a dark blue French Provincial look I'm guessing but its just a guess. And the bed had a nice scallop in the middle which forced the body to roll back to center every time. Safety first I guess.

I opened my duffel bag and found that I needed to do some ironing, and since I had a couple of hours to kill in this glorious room, I was thrilled to have an activity. I looked in the closet where in most business oriented hotels one would find an ironing board and an iron but not here. Figuring I had two things to order from the hotel, food and the ironing board, I ordered the food, then I ordered the ironing board. Now get this, housekeeping informs me that I can only have the iron and the board for 30 minutes. "Why" I asked. "are you busy right now?" "No" she said explaining that it was just the hotel's policy. "Do you want it now" she asked, "Not really, I have food coming now" I replied. I'm sorry but that was a first for me. "Yes we have an ironing board, but you can't really use it".

Remember chrome/foil wallpaper? You know, intensely reflective wallpaper with some ink pattern printed on top of the foil to make it look "cool?" Close your eyes for a moment, no wait a minute then you can't read, OK open your eyes and try to picture powder blue foil wallpaper, with really intense crystal light fixtures in the bathroom. Got it? Its true. No lie. I have pictures. That's enough ripping on this hotel, and no I do not recommend it when visiting Manhattan. Do not be fooled by the attractive room rate. Spend the extra $40 per night and stay somewhere else. Of course this is only my opinion, do as you will.

Don the Global Traveler

In between the hotel and apartment shuffle for those few months in Manhattan, I met and made some pretty close friends there. My opinion started to change about this town because the people, once you got to know them, were pretty generous and warm. This is also where I met the one of the most energetic and dynamic people on the face of the Earth, Don. While I was taking care of the architecture required to support the new EPM 8.0 product, Don was heading up the report build team for PeopleSoft. Since the product was still "Beta", there were a few kinks to work out, but they all were in pretty close communication with the Pleasanton development community and getting problems fixed as they went along.

Don and I couldn't be more different and that statement could be made on many levels. Don was a runner and in great shape, I wasn't. Don was a handsome man, I was well.... a little taller that him. Don was single and living the fun life often traveling the globe, I was married with 3 kids.

Somehow though, we fed off each other when we went out for dinner or drinks after a long day at work. New Yorkers like to work hard in the hours they were actually in the office, which left plenty of follow up work for the traveling consultants to do after they locals all went home for the day. Late as most nights were, Don and I still found time to eat and experience a little culture on that overcrowded island.

The real restaurant jewels in Mid Town are the ones down the side streets between 40th and 60th . Little family owned Italian, Cuban, Japanese restaurants, they were all wonderful. You still need to be careful and check the menu on the outside of the building. Any restaurant that won't post their bill of fare is suspect. Some New Yorkers argue that the "best" restaurants wouldn't dare post their menu. I argue that the New Yorkers mean "fashionable" not "best.

Even with those guidelines and your own cautious nature, there are still ways for a restaurant to quietly get your money. Don and I visited one of these little out of the way Italian places without a menu posted outside. We behaved like tourists and asked to see the menu, but the prices were tolerable, $19 to $35 for entrees a la carte.

We were seated near the rear of the restaurant practically alone when the waiters suddenly became very attentive. A green colored bottle of water was opened, not at our request mind you, and the glasses filled. The water was perfectly chilled and tasted wonderful but I did not recognize the label nor could I translate the Italian description. Just then a wonderful plate of Antipasti was placed on our table along with fresh bread, olive oil and balsamic vinaigrette. "Who needs a menu" I blurted.

We eventually ordered our entrees, and a $35 bottle of wine. The meal was wonderful, and memorable. It was memorable because after I got the bill, I practically swallowed my tongue. Remember all that wonderful water, and bread and Antipasti? Don and I had each ordered an entrée, and dessert (that part was our fault) but the check came to $240, not including tip.

Apparently everything was a chargeable item including the bread and we just kept on eating and drinking. I think the green bottles of water were $11.50 each. The Host who greeted us at the door initially, offered us big traditional Italian hugs as we departed.... Maybe he was checking me for my wallet, maybe he was gay, I'm not sure.

Many nights were spent at the "W" Hotel in the bar affectionately known as the "Whiskey Blue". If you were a hotel guest you could easily enter, but all others were stopped a the door and evaluated for the proper amount of black clothing. If you were mostly dressed in black but with shocking color like "charcoal grey" mixed in, you might be suspect and denied entry.

Don and I always brought black and made it our regular hang out for a while mostly because of the great people watching. The women were models, and most of the guys seemed to fit the wall street pit-bull stereotype. Ah well, the domestic beers were only $7.50... HOW MUCH? I started to miss Lacrosse Wisconsin and the 10 cent shells.

The work continued in Manhattan for about 6 months and then it was time to hand off the project to the customer for completion. I did enjoy my time in Mid Town despite all the challenges of living there. At first glance New Yorkers sometimes seem to be a cold people, but that's only an external shell.

Once you get to know them, work with them, get past the big city defensive exterior, they are great bunch of people. They put up with a lot of crap to live and work here and that gives them plenty of character. The New Yorkers I met with families have moved out of the city to more suburban/single family home areas, but then have to suffer ridiculous one or two hour commutes just to get to work. I mean you gotta love this town to stay and most of the folks here do love their town.

From a visitor's perspective, it's a fabulous place to spend some time, because it's a visitor's town. The shopping is fashionable and fabulous, the food here is fantastic and the entertainment is endless.

That's it for alliteration and assonance. Don and I parted friends and we vowed to get back together again soon.

Back in Detroit, the changes to the management structure continued under the Conway reign. Some of our senior management team chose to move on as soon as Dave Duffield handed over the COO responsibility, and others were brought in to the company to become part of Craig's core team. It was hard to keep up with the changes after a while, and the farewell notes kept coming from many of those who helped build the company. Baer Tierkel and Peter Gassner hung on for a while, and took on more marketing duties as the news of PeopleSoft 8 started to roll out to the public.

The whole of the year 1999 was fraught with changes for PeopleSoft. January of 1999 was the first time in PeopleSoft's history where cross region and department layoffs were planned and executed. Some corporate departments were consolidated or reduced in size and the consulting ranks were hit hard as well. Anyone not assigned to a significant project, or billing on a regular basis was shown the door right after the holidays. Some 430 employees in all. Many of the consultants let go were friends.

Through 1998 and into 1999, the PeopleSoft stock price fell from a peak of $56 to $16 in the span of a few months and followed the rest of the tech-stock market in a meteoric slide. Some of the employees cashed in at the right time, but most of the employee base held on to their shares and options looking more towards retirement rather than that big house on the hill. In retrospect, the big house on the hill approach would at least provided us some real-estate rather then deflated stock options that sat under water.

The field organization kept rolling on, focusing on customer issues and projects like we always did. The push for Y2K upgrades and installs was still pretty strong and most of the remaining staff were busy helping customers get new product running in production or older product upgraded before the end of the year.

Other regions were starting to experience a different kind of demand from customers though, and it involved a more integrated approach to building their ERP solutions. There was a constant shortage of specialty technical resources across the country, but in addition there were a limited few who could bring all the architecture together in a single picture and talk about the solution. We found ourselves learning a different kind of consulting, at a higher level of the architecture. It was not only important to know the whole technical picture but also be able to translate it in plain English to senior management.

Chapter Thirteen – Growth and Changes

The end of 1999 proved to be a time of changes indeed. Our customer base had grown quite large and despite some of the WAN latency challenges of using our software over large distances, PeopleSoft still had a respectable number of customers with global solutions in place.

The new PeopleSoft 8 product release promised to reduce the amount of network traffic significantly but many of those technical details had not been released to the public. Our little advanced technical team was well aware of the current trouble being experienced in the marketplace.

There were plenty of "technical" people out there, but few who could put the whole puzzle together. One of my compatriots who used to manage the install consulting group, was planning to build a new team who would focus on the higher level view of technology. Not just providing people who knew how to code, or people who could tune a data base, but people who knew enough about server architecture, tuning, networks. Consultants with experience in building a development environment and a disaster recovery solution to boot. They would need to consult with higher levels of customer management to do the planning and strategy for the whole project. I liked the sound of it all.

Unofficially we called ourselves called "Technology Solution Delivery Managers" but were listed on the HR books as "Regional Service Line Managers". It didn't matter, we were starting to develop job descriptions that would fill the void in our consulting ranks, and provide a service to customers where we could help set the right course of action rather than help fix things when they went wrong. The concept was one of proactive strategic consulting and we were all "geeked" for lack of a better term, about the chance for the groups success. Once again the Midwest region was leading the way on the technology front.

Four people were picked for the fledgling group, each of us having a different focus on technology but all of us experienced in presentation work and working with senior management teams. One focused on portals and web technology, another was upgrades, a third was database architecture, and the fourth was an architecture generalist. The fourth one selected was me and although I was light when it came to depth in a particular piece of technology, I had more experience on putting together the whole picture from interfaces to data centers.

Perfect as the picture seemed to be, there was always the chance that nobody else inside our organization would know what to do with us. How do you present the concept of a more expensive "consultant" to the customer. The competition was increasing from firms that PeopleSoft had to call "partners" and it became a war of cheaper consulting rates when the resumes of candidates looked so similar. "Similar?". It seemed clearer to us who exactly had the edge in training and experience… but. Who worked for the software vendor and who didn't? Who had to accept responsibility for the software and who didn't?

In many cases it was all about the money. If you could come up with a cheaper proposal for resources you were the winner, regardless of the quality of the resources in question. When projects started to go astray, our group was often called in to do an audit or project evaluation. When we found that the root cause was poor consulting by our "partners", we often had to downplay the findings in order to maintain the "relationship". Very often our people were subcontracted to a partner so that they could mitigate the risk of the partner led implementation. Of course the partner still led the project, so by association they deserved the credit for keeping the project out of trouble. Sure. We all felt good about that.

These were turbulent times for the consulting trade. Due to the tightening of the stock market, customers started to get frugal with expenditures of any kind. Aside from the investment they had to make to prepare for Y2K, there was not much else in the way of new product going live.

Despite the sales team struggling with how to position our skills or what benefit we could provide the customer, we prevailed by attending customer meetings and pre-sales events. All we needed was a chance to talk about the complexity of a global solution and how we could help them prepare for a successful implementation.

All we needed was a voice at the conference table. It was shocking really how many fortune 500 companies we visited, where the technical people in charge had no clue on how to bring the whole (integrated web, application server and database server, network and security) picture together.

The news of our success spread to other regions albeit slowly. Stephan who was leading our team, provided oversight and the other regions began to form their own Solution Delivery groups. Each team seemed to prove their worth inside their own region, to their own management team. It was a solution that would focus on that particular part of the country, that particular set of sales people and project managers and that particular set of customers. Though Stephan and the rest of us started to set standards for "technology solution delivery", it was up to each region to implement their own standards and processes to fit their needs.

Vacations were on hold at the very end of 1999 and we had been working with a number of our customers testing systems for the year 2000 changeover. Most all of the testing yielded no anomalies and we felt pretty good about the success of our software in the market during changeover. January first came and went without the power grid failures and computer systems going down that doomsday predictors said were inevitable. Most people will say anything on camera to get their 15 minutes.

Things went smooth across the board so much so that certain companies started to wonder if there was any risk at all. Yes, we all thought. "But it was all the investment in testing and corrections to the software that mitigated the risk". "It went smoothly because we beat the problem into submission." You just can't make people happy sometimes.

With the new job I was spending more time at home which made everyone happy I think. Me for sure, Cheryl for sure, the kids I'm not so sure. I was just another parent around to make sure they weren't getting away with the things they used to. OK. Not necessarily true. I got more hugs and more stories from school and was able to see more of their activities. The girls had both enrolled in dance class and my son was progressing beyond the local "house" hockey program and he was taken aboard a travel team in the Spring. It was nice to get a chance to some of their activities and cheer them on from the stands.

The new job didn't require us to be 100% billable on the road. Because of our skill sets, we were also valuable in the office. Working on proposals, helping with staffing models for customers, chairing conference calls and participating in pre-sales efforts were all part of the new gig.

Even if it proved to be a long day, we could still be home to sleep in our own bed. There was still the project management side of the story however. Sometimes we would be the senior ranking technology person available and by default we would have to manage the team of technology people on the ground.

For some of the larger projects active after 2000, the managers in our group took on more and more project management responsibilities. Even the executive management team in Pleasanton was pushing for us to be more of a project manager, than advisor and architect. We were scheduled for mandatory project management training and reluctantly we all went through the classes offered by ESI and George Washington University.

This shift in direction from a technology focus to project management had our team talking frequently on the future direction of consulting and of course our own careers. Where does a consultant go after being declared a "senior" systems analyst. What management opportunities are there for a technology person. Are they then limited to project management? It required an entirely different set of skills and training, where very little "technology" was ever called upon to do the job.

It was a direction we didn't understand because we had proven that there was a new role that worked at a customer site. One of trusted global architecture advisor. This was a role that we had developed while at customer sites when introduced to the right level of senior management. If we could get near the top, it was easy to talk about other customer "best" practice, how to manage data center consolidation efforts, how to self -insure disaster recovery practices. Our whole group was very successful in building the "trusted partner" relationship because we most often knew more about the marketplace and technology that the customer did. Not because we were smarter than them, but because we were exposed to more customer solutions, more different types of technology, we had the field experience on PeopleSoft 8 which by definition was the cutting edge at the time.

PeopleSoft 8 was the first enterprise wide software suite, totally based on the internet for delivery of the solution to the desktop. Taking advantage of current relationship with BEA, we made the application server middle tier more powerful and it managed not only integration traffic, but pure XML messages along with our own HTML pages from PeopleSoft. The web server was now an integrated component of our architecture and it changed how customers had to think of doing business in the "Internet" age.

Again the "Technology Solution Delivery Group" was busy working with customers on how to set up the new architecture without trashing the investment they already made for the 7.5 product. They

131

were pleased in the end, that they could still use their old server components in the new architecture. We liked making people happy.

There were challenges at some larger customers however. For many of the Fortune 100, there was a significant investment made within their own IT department and technology specialty teams. Without mentioning names, at least three companies come to mind where it was impossible to influence their technical management team. These were managers who had been identified by their company as "up and comers", on the "fast track" to promotion and responsibility. There were actual HR programs used at these companies to track and promote these employees separately. These were typically the folks in charge of any technology group at the customer site.

There wasn't really any point to arguing who was smarter, because they were and they knew it. In fact one guy made it a point to tell me about his educational background and awards. "That's great" I replied. "How much experience do you have building distributed global database solutions?" I would then ask.

It was a challenge forming any relationships at certain customers because of their management infrastructure and the constant filtering of any recommendations that came from our team. We tried to help them and that was about all we could do. Ultimately, they did things their own way, morphing many of our designs and ideas into their own which I guess in retrospect, was OK too.

A Mid Year break

Around July, the consulting organization sponsored what would be its last international consulting conference. Bombarded with memos and voice mail messages about how this conference would be used as a serious platform for training on PeopleTools 8, and its new upcoming features, we wondered why they picked Las Vegas for the location.

One e-mail warned us that we would be required at all sessions and (like grade school) attendance would be taken. Ah, yes. It was more fun in the old days when you were responsible for your own behavior but also ran the risk of being publicly humiliated by management if you missed a meeting. Las Vegas. No distractions, just pure educational focus in an environment conducive for learning. Yeah sure.

Now mind you, I've been to Vegas before, so the scene on the plane and at the airport was familiar but still disturbing just the same. This particular Airbus A-320 was filled with retired folk, some in wheel chairs, some with walkers, all on their way to spend what I'm assuming were their retirement fund distributions, or social security checks. One tends to make this observation after looking at their clothing and the quality of their dental health, when they would yawn anyway. It didn't appear to me that these folks were on their way to joyfully piss away recent lotto winnings or 'extra' inheritance money.

The air was smooth and the flight attendant service in the back of the plane was all but non-existent. It was a familiar story. I had accumulated so many flight miles the previous year, I was getting quite used to the free first class upgrades, but not on this flight. I had gotten spoiled, and soft and used to the special treatment in previous months, and the endless supply of Bailey's on the rock was pretty nice too. Even on those short flights without actual food at your disposal, there was always the Bailey's.

We had a pretty good view of the Grand Canyon as we flew over, and I remembered that with all my east/west coast travel, I had never really gotten a good view of the canyon until that day. For lack of a better word, It was "grand". Our approach to Vegas was smooth, the ear popping minimal, and the landing could be categorized as a two hopper, but we arrived at the gate without injury. That alone is my measurement of a successful flight.

Grabbing my featureless ballistic nylon duffel, I made my way through the dead air stench of a 4 hour flight cabin and headed for the terminal tram and eventually the airport shuttles. The PeopleSoft logo could be seen throughout the terminal as throngs of FNG's with bright blue computer bags followed each other in the hope someone knew where they were going. They looked like cattle herded in a large pen waiting for the gates to be opened to the slaughterhouse. Following each other to the one exit identified for their shuttle use, the throng politely waited in line for the kill, er, I mean the trip to the hotel.

It was funny how the exit to the left of the crowd was free of any line, and for 6 bucks, I could get an immediate shuttle ride to my hotel. Hmmm, I guess I was thinking outside of that box again. I paid the 6 bucks and enjoyed the air conditioning while the FNG throng waited and sweated.

Here's an acronym checkpoint. FNG = Freaking New Guy

Assumptions and Disappointments

My shuttle driver was a tall and articulate black man about 35, from Jamaica I was guessing based on the accent. He was friendly and quick to strike up a conversation on our short little ride to the hotel. He asked me if I was in town for the convention, and I replied in the affirmative. He then started to talk about all the people he was driving to the Hard Rock Hotel, and specifically commented on the women.

"You wouldn't believe some of them" he said.

"And the clothes, I've never seen so much spandex".

At this point I gathered he thought he was taking me to the Hard Rock Hotel so I told him that I was actually attending a convention at the Mandalay Bay.

"Oh, I thought you were here for the porno convention like all the rest.." he said.

Totally in reactive mode, and not really thinking about what I was doing, I looked down into my lap and said

"No, I wouldn't be here for the porno convention".

My shuttle driver just laughed, and I slumped down into my seat and then we talked a little about the exciting, and glamorous world of software consulting. I remember tipping him especially well, maybe it was because he didn't laugh too much when he found out I wasn't in the porn trade, maybe it was because he actually thought I was.

Mandalay Bay is as close as you can get to a family oriented hotel in Vegas, although you just have to appreciate the oxymoron there. It hosts a fabulous bunch of restaurants, and a great outdoor wave pool and play area for kids, close to the MGM Grand where you can hit the mini theme park with coasters and other rides.

You never forget of course that the casino is right downstairs and porno convention is down the street. I checked in, and headed to my room to dump off my stuff. Carpet is carpet and curtains are curtains when it comes to a hotel, but the bathroom, now that was impressive. Floored in Italian Marble with a long counter and dual sinks, it boasted a huge jetted bathtub big enough for two, and a large corner shower, also big enough for two. It was the kind of bathroom you might see attendees from the other convention use in one of their films. The fixtures were glamorous, and the mirror was generous and the mind started to imagine all sorts of evil little vignettes taking place in this rumpus, er... I mean bathroom. Suddenly I was afraid to touch the countertop.

My roommate and old Chicago buddy Jason had not checked in yet so I retreated to the lobby to look for some familiar road warriors and was lucky enough to spot Amy, another warrior buddy from Milwaukee, just checking in. We exchanged our usual wide smile greetings and after dumping off her stuff, we made a trek our into the "concrete and neon" jungle.

Some folk tend to think of Time Square when they hear that description but I assure you that Manhattan has nothing on Vegas when it comes to lechery and treachery per city block, to say nothing of the volume of concrete that exists there. In the length of a single city block I had been handed at least 15 separate business cards or brochures with full color pictures and graphic descriptions of what "Candy" or "Cynthia" or "Roxanne" could do for me in the privacy of my hotel room. It was "Only $95 for the first half hour", at least that's what the business card said, and I found myself doing a little international price comparison.

You see, in Amsterdam the price in the red light district depends on a number of factors, or so we were informed by Piet the previous year. The number of picture windows lit up, or the number of girls working, the activity in the streets, weekdays as opposed to weekends all affect the going rate. From what Piet told us, you can get a little government regulated booty for around 60 Guilders or $30 when the market is down. If you're a bargain hunter, any Wednesday in the tourist off-season is a good day for booty in Amsterdam. <Ahem> I digress. So American pricing is a bit higher in general, but then again that KLM round trip ticket to the Netherlands kind of offsets the cheaper services, as it were.

Losing a Quarter of a Million Dollars

Amy had enough of my fascination with the street literature, although it was fine reading in its own right, and she suggested we hit one of the Casinos and check it out. We took the overhead complimentary railway to the MGM Grand and walked around quite a bit, not realizing the scope of the Casino's square footage. Off in the distance, next to the one of the narrow pathways was the "Elvis" slot machine. It was emblazoned with an image of the "King" with the big hair and the white studded jump suit, a rolling jackpot amount on the top, and it called to me from the distance.

I told Amy that I just had to play this one, after all it was "Elvis" for gosh sakes. I fumbled around my pockets for a couple of dollars and built up my credits to 8, just for the "King". My gambling prowess was non-existent though I wanted to follow some logic when playing this machine, you know, one quarter at a time when you're losing, then the maximum credits after a few losses in the hopes that you'll hit a winner. This worked too, as I hit a winner for 15 credits (quarters) so I kept on this logic path for a few more pulls. After I hit another small payoff, I went back to betting a single quarter. Just then my machine came up with three Elvis logos across the payoff bar. I pointed at the machine and looked at Amy who promptly shook her head indicating she had no idea what I was pointing at.

Two chairs down from me a Japanese woman started screaming something at me that sounded something like "Yoo Waa" "Yoo Waa". Of course I thought she was nuts, but then she pointed up at the jackpot sign and kept repeating the "Yoo Waa" theme for what seemed like an eternity. OK, I got it. "You Won". I then saw on the machine that if I had bet the maximum, I would have won the jackpot. But, I had just won about 60 dollars so I was thrilled.

To me it was no big deal, and I was only there to play the machine because Elvis was on it, I never actually expected to win anything, and I don't really gamble. It seemed to be a more crushing blow to my Japanese friend than to me, and I fully expected to toss a few quarters at the feet of the "King" just because it seemed to be the thing to do. I mean, while in Vegas, do as the Vegas-ites or Vegas-ans do, right?

We turned away from the machine, though I looked back to see what the jackpot actually was just in case I decided to write a story about it someday, and it cashed in at 213,714 US Dollars. To add insult to injury, another woman with ragged hair, second hand clothes and a cigarette hanging out of her mouth stopped me to ask me "You didn't bet the maximum, did you?" her voice rose a ragged octave "YOU ALWAYS BET THE MAXIMUM"…….We kept walking. "Dummy" I thought I heard her say.

Now I usually don't let gambling addicts with tattered clothes get to me, and this one didn't, but friends and co-workers are a different story. I jokingly told the story to a couple of buddies back at the corporate reception and we all laughed about it, after all it is pretty funny.

It seemed to spread from there, the story taking on a life force of its own, some guy losing $213,714 on the Elvis Slots. The next evening I met a few new folks for the first time as they were from a different region, and during the conversation we talked about losing money in

Vegas. Some of the people in the group had lost a couple of hundred, others were up a hundred or so.

I commented that I had only spent about $8 so far on the slots and that was it, when one of the recently acquainted started to say, "At least we're not the guy that lost the quarter million on the Elvis machine."

Everyone laughed. So did I. Then I added "You always bet the maximum... even I know that", then I quickly walked away.

I later ran into some people that I knew, "Thank God", and we stopped for beers, margaritas, and a rum punch or two, and had a great time just catching up on each other's exploits the past year. Somewhere during the exchange we switched into story telling mode and swapped tales from the strange world that is software consulting. I felt I could offer the Elvis story for a quick laugh, but I didn't expect the laugh to be so damn long. Some of them had tears in their eyes. One of them asked "That was you?" "The story's all over the conference".

"Sir, I knew Andy Warhol, Andy Warhol was a friend of mine, and you're no Andy Warhol".

So this was my fifteen minutes of fame eh? It seems that my situation and my notoriety didn't really fit the spirit of the Andy's famous quote, I mean after all, it shouldn't really qualify as fame if just 3000 people knew right? I convinced myself that this was no Andy Warhol situation at all, this was just "Roadkill Fascination". Maybe that can be my quote instead of Andy's, and my dilemma as well. People will slow down and stare but in the end they just drive by. Roadkill Fascination. Kinda catchy ain't it? Except that I would be the Roadkill in this example, whatever.

A small group of us had left the corporate festivities and made our way to a Café attached to the Paris Hotel. We bumped into Janice and pulled her along as well. From there we could watch the dancing fountain show at Bellagio across the street and the location in general was pretty cool. We had good wine, good people, good view of the strip and some of the best people watching in the world. From that café you could see them all, the rich and poor, the beautiful and the not so beautiful, the young and the old, and the customers with their personal service providers.

On the long walk back to the hotel, I mentioned that I really needed a chocolate chip cookie and a big chocolate milk. As we passed through one hotel Janice said "Hey, what about a doughnut?" "We just passed Krispy Kreme in the last hotel". "That'll do" I replied and we went off to scam a half dozen or so awesome doughnuts for late night snacks. As luck would have it, they had chocolate milk too. Yeah Baby, I was a happy man. Janice knows all the best places, hands down. On the way back, I received pretty large cash offers for the Krispy Kremes in the box, funny how people will pay just about anything for the right commodity at the right time, but I said "no". I might have been willing to negotiate an exchange of commodities with the right service provider, but of course, that had to remain a fantasy. I returned to my hotel room, I munched, I showered, I slept.

All strange things must come to an end, and so did the Vegas Conference in July. I left Elvis, and the porno convention behind to return to the sanity of Midwestern America. I got home late on Sunday and had to fly out Monday morning at 7 AM to a new customer who had some of our software people already on site. When I arrived at the customer later on Monday, I was introduced to a number of the customer's project team, one of which had been briefed in advance of my arrival. I said "Glad to meet you" and his excited response was "Are you the triple Elvis guy?"

I guess Warhol was right after all.

Chapter Fourteen - Millennium

Towards the end of 2000 and the beginning of 2001, a number of things were percolating at work and at home. Our team continued to grow in demand and we had identified a few more candidates for promotion into the group. Around about the same time, Stephan and I were talking about some of the issues the other regions were experiencing in building their own technical solution delivery team.

Not so much a situation where they didn't have good people to build on, but each region was left to use the Technical SD team to do with as they pleased. The mission statement for each regional team was different, and the regional SD teams were doing more project work than technical leadership work.

We weren't sure if it was a situation where the sales people just didn't know how to sell the concept of a strategic technical team, or if the demands of the specific region required more project managers than technical leadership. In any case the "shape" of the regional technology solution delivery teams fell out of alignment with other regional teams. Stephan and I along with others saw the need for a centralized mission statement and a common set of goals for our most senior technical teams.

We felt strongly that when it came to advanced architecture issues, system design and problem solving tools, that the message from each individual region should be at least consistent. We all thought that the team structure needed to be standardized as well, providing specialty resources to each region when needed but also providing a promotion path for those in the consulting ranks ready to accept more responsibility.

Ideas and Punishments

Stephan and I started to put together a presentation outlining these thoughts and recommending that we make the Technology Solution Delivery group a national entity rather than regional. We would never abandon the idea of regional coverage, but with a national voice in place, we could help shape the mission, the tool set available, the training, and the consistency of our recommendations to customers. Sounded more than logical to me.

From my perspective this kind of thinking, I mean trying to come up with ways to improve customer service and standardize a technology vision for customers, was what we were supposed to do. I imagined that a company specializing in higher technology and advanced software solutions, wanted its employees to think outside of the box and suggest alternative ways of doing business. I was apparently wrong.

Perhaps we needed to chalk it up to picking the wrong time to make the presentation. The consulting organization was still being re-organized and new people were being moved into leadership positions out in Pleasanton. Who knows. What we do know is that Stephan made the effort to present his ideas and took the brunt of the negative feedback in return.

The corporate senior management team made it sound like we were trying to take roles and responsibilities away from other departments. "Build our own fiefdom" was one comment.

"No, you missed the point".

"We're trying to improve the way our own group does business, standardize procedures and best practices, not take away somebody else's job."

It was a disaster. The whole idea was quickly buried and Stephan felt like he had exposed himself so badly that he was looking for management opportunities out of the consulting group altogether. I just love negative reinforcement as a means to improve productivity.

In the spring of 2000 I had my annual performance evaluation from Stephan at of all places, the Original Pancake House. Stephan drove in from Atlanta and met me on a Saturday on his way to borrow my family's cabin in Northern Michigan. He had brought his dogs along and planned to hike the trails in the National Forest.

We met and it was a glowing review. Then he laid the shocking part on me in between bites of omelet. Stephan had discussed a transition plan for himself with the regional management and my name was offered to them as the prime candidate for Director. I swallowed hard.

"How did they respond to that?" I asked.

"They were pleased and agreed with my recommendation." Stephan replied.

"Does this mean you're leaving?"

"Not right away" Stephan replied then went on to describe how he felt it was best to move to another area of the company offering him better opportunity.

I was sad for Stephan but knew he would wind up on top again, and flattered that I was thought of this well in the region. All in all, it was a fine breakfast.

The Pay back, June 2001

That summer after saving some money during the previous year, I paid my wife back for the earlier business trips I got to take to Europe without her. For our anniversary that year I suggested that she pick a country she would like to see and I would make all the arrangements. I hoped she might pick a country where I could speak a little of the language, but it really didn't matter. She chose Italy, and left the travel itinerary up to me.

I didn't know a whole lot about the country but after a little investigation, I settled for the Northern district on Lake Como. Working with a local travel agent we found a beautiful turn of the century hotel right on the lake, in the middle of a small town near the Swiss border. It was as it turns out, where the Italians like to go on their own Summer vacations.

With the flight miles I had accumulated, it was easy to pull first class tickets on the upper deck of a 747 and that made all the difference in terms or arriving rested, or arriving frazzled. I didn't want to know how much the tickets would have been if we had to pay for them. We connected into Milan and found the Hertz counter without any trouble and we tried to get used to the fact that there were military people everywhere with machine guns on their backs. I remember thinking to myself then that it was a good thing none of those terrorist threats existed in our own country.

We had a ball driving on the Autostrada in our little Opel minivan, and tried to keep up with traffic where the 140 KPH speed limit was merely a suggestion. German sedans ruled the left lane but we made it to our little town of "Tremezzo" in due time. From there we took day trips to Venice, Pisa, and two trips to the Alps through the small lake towns of Switzerland. After almost a week of no distinct plans but great times, we boarded a train to Rome and finished our trip with a few days in Rome just off the Piazza Navona, near the fountain of the Four Rivers.

We talked at length for the first time in years mostly because it was the first time we had taken such a trip without the kids. Cheryl's parents took over kid-watching duty and it was nice to get to know each other again. We talked about kids, college, the next house, money, and my job.

We especially enjoyed the European custom of the 3 hour dinner. Hmmm, socialization and a relaxing dining experience... what a concept. As far as the job was concerned, we tried to come to some sort of agreement on what things worked and what things didn't. In the end, although she would prefer me to be home all the time, it was better now that my current job allowed me to be home some of the time.

September 11, 2001

Most everyone I talk with knows exactly where they were the morning of 9/11. I was in the Detroit office with a small number of sales people and developers when someone down the hall yelled something about a plane crash in New York. I remember looking on the internet for news reports first, but finding nothing. A few seconds later, one of the developers produced a little black and white travel TV and plugged it in so we could see what was going on. There it was, the WTC tower with smoke pouring out of multiple floors.

It was sort of strange watching it in black and white on a 5 inch screen, looking almost like a 60's period piece on New York, except this was live. Then while we tried to make sense of what idiot pilot could have made this kind of error, we watched in horror as a fast moving image emerged from the side of the screen only for a split second before plunging into the second tower. We knew then exactly what was going on, but couldn't believe it. "Oh my God" were the only words spoken.

News reports started speculating about other targets, then broke the news about the Pentagon, it was then that we heard that all air traffic was grounded by the FAA.

I went back to my shared office space and realized that our building, was one of the tallest buildings in the Detroit area, sitting on the end of a cluster of buildings in Southfield, a suburb of Detroit. I looked out the window towards the Detroit International Airport and I could see planes in the air. I watched the planes for about 15 minutes still numb from the news.

Suddenly the phones started to ring and local managers and directors were actively trying to account for every employee out on the field.

Lots of people were on planes today, because that is what we did for a living. I had no active projects that week and nobody that I knew of was traveling that day, but we all pitched in to try to locate everyone else.

It was the first time in a while where I felt like the company I worked for, cared about the people who worked here. We didn't stop until everyone was accounted for. Luckily, everyone survived. There was immediate concern for some people from our office who were supposed to be in Tower One that week, but we found that their meetings were thankfully moved to another city.

The year rolled on as we tried to keep the business running after the September kick in the teeth, and I was performing the role of auditor more than anything else. In most cases I was called on to evaluate an architecture or a project plan and find out where things went wrong. I never seemed to be invited by the customer, but was called on by our own management team to provide a report back to them. In one case things were so poorly managed by one customer that the report was extremely critical of the customer's management, the partner responsible for the implementation, and the competence of the customer's own technical people. I tried, but there was no way to "put lipstick on the pig".

Since I was the only PeopleSoft manager on-site, I stood in the CIO's office while he read the report over the phone and our management listened in via conference call. His face turned red and he began yelling into the phone.

"How dare we turn in a report like this?" he went on.

When the call was over, I told him I was willing to discuss what I found and explain any part of the report he liked. He stood up at his desk and threw the 40 page report at me. It hit my shins and landed at my feet. I refused to move. Next this 6 foot 8 inch former college basketball player barged towards me bumping my left shoulder as he stormed out of his own office. I waited there willing to take on a discussion or an argument or whatever, but he didn't come back.

I even asked his secretary if I should wait there, she said "she didn't know". That was a fun day.

Other customers were more appreciative of that type of audit, because the idea was to correct things that were wrong on the project. Things continued this way until the end of 2000 and our little group continued its travels past the boundaries of the region helping out other areas of the country as needed.

Internal E-Mails were strangely quiet in November for some reason and I heard a little back office talk that there might be a position opening for Director in Solution Delivery. At one point I was asked to interview a guy from one of our partners as a possible candidate for a manager position in our group, but that was it for "interview" opportunities in the region. Nothing else was posted.

I wound up talking to this new guy named Matthew and found him to be quite nice, with a sense of humor and a great technical background as well. We talked quite a bit about the region's customer base and the kind of work we would be doing in the Tech Solution Delivery group. All in all it was a good interview and I liked pretty much everything this guy had to say.

A couple of weeks later the mystery was solved and the new guy, Matthew, was named as the new Director of Midwest Technology Solution Delivery where my old friend and boss Stephan had taken a position in Europe. I also expected it was to get out from under the management team that hammered him for his ideas earlier in the year.

The shocker for me was that I never got a chance to interview for the Director position, after hearing that the regional management team thought well of me in that slot. Somehow, they chose to go outside the company for a resource and the position was never posted internally for interview.

For about two months, I was crushed. I wasn't sure how to interpret the whole process or lack of process, and I tried to find explanations that reassured me about my own worth to the company. Or more to the point, did I have any worth at all?

Stephan suggested that I confront the VP in the region to ask about the rational for the decision, and I did just that. Confront was a strong word and in the end it was more of a friendly meeting than a confrontation. The VP went on to list the background she as looking for in a Director, including managerial skills, marketing skills, and the ability to build a practice from the ground up.

I politely asked if she had even taken a look at my resume, because everything she was asking for was in it. In the end it was her decision and she was confident that "Matthew would be perfect for the role and the decision should in no way reflect poorly on my value to the region". OK, at least I got a chance to ask.

After some sort of reflection involving Canadian Beer and a couple of days off I realized that it was still good to be home more with the Solution Delivery group. Assignments had me jumping from state to state but there were bigger breaks between assignments as well. I did spend some time with a customer in Omaha and enjoyed working for Midwestern people again.

Back to the Heartland

Omaha is a medium sized agricultural town struggling to build a modern identity. You can see it in the revitalized downtown area where storage buildings and warehouses have been converted into restaurants and shops. The restaurants are good, and you might say even bordering on "trendy" in certain cases. It is a pleasant walk around the new area and a few new hotels have popped up to support some new business activity in the area as well.

A few of our technology brethren had been assigned to this one customer in Omaha for a couple of weeks of analysis. One memorable evening out after a 10 hour work day, three of us had decided on Mexican food for the ethnic dinner choice. We found a little place recommended by the customer manager we had been working with, down a flight of stairs below street level. Paul and Rajesh liked the menu and we all agreed on chips and salsa as a starter.

"Interesting choice of salsa dispensers" I thought while looking at a squeeze catsup bottle with the tip cut off.

It was a challenge to get the salsa out of the bottle and the thicker tomato chunks would clog the tip. Rajesh grabbed the bottle and pounded it on the table to free the clog then proceeded to grab it with both hands.

Perhaps the worst choice of the day was deciding to wear the white shirt, or maybe it was choosing the seat against the restaurant wall across from Rajesh. The salsa created a wonderful outline of my shoulder and neck on the back painted wall as it sprayed evenly over the left side of my white shirt. I sat in shock for a moment and Rajesh remained silent as well, maybe waiting for my reaction, I'm not sure. As the dinner was brought to the table we all were laughing about it, but I was the one who would later be washing clothes in the sink with shampoo. Rajesh soon left the country and migrated to Canada, never to return.

The following week a larger group of us including some of the functional product experts wanted to take a break from the week and we all decided to catch a baseball game. Omaha is home to the college world series at the Rosenblatt Stadium and also home to the Omaha Golden Spikes minor league baseball team. We sat in the sun and ate ourselves into a coma. Hot dogs were two bucks, ice cream was served in plastic helmets, and the beer was cheap. Oh yeah, and the baseball was fun to watch too.

It seems that Midwestern folk are pretty easy going (Except for Chicago), but they can stay focused on the work that needs to be done. Not that Chicago people can't be easy going that is, but I get the impression that some of them are really frustrated New Yorkers who are trying to work more hours, drive faster, and live faster than their counterparts in NYC. It seems to be a case of Metropolis Envy. "Oh yeah, well we have skyscrapers in Chicago too, and I have a BMW, and I hang out at the "W" after work too."

OK, I exaggerate but it feels like that when I meet some of them. The work ethic fits here too. New Yorkers like to work fast and hard mostly because their commute is so freaking long that there is only about 6 hours a day to get anything done. Then they drive like bats out of hell to get to the bridge or tunnel, or ferry, or wherever. That describes some single or married without children professionals I've met in Chicago. Good people all, but get them in their Range Rover at 5PM downtown, when they have to be in Oakbrook at 5:30 for dinner…… I've seen them drive on the shoulder and run Geo Prism's off the road. "Get the frick out of my way, my Appletini is getting warm!"

I spent very little time in Chicago that year because of an expense crunch we would experience for about 2 consecutive years. Not that we knew that then of course. The stock market was down, and everyone on the technology sector was taking a bad beating. More dot-coms that held out after the first round of bankruptcies, finally succumbed to the marketplace and went under. Even the software sales pipeline narrowed, and if you were to survive, you had to take a competitor's dollars because "new" software money just wasn't there. It was a brutal software sales environment.

PeopleSoft still did well that year still posting total revenue in excess of 2 Billion, and the profit margin stayed respectable because of expansive cost containment initiatives. On the consulting side "cost containment" meant staff reductions.

For those people "on the bench" or not currently billable to a customer, continued life at PeopleSoft was a tenuous proposition. The regions were made responsible to take their own action with regard to salary expense and whether or not they would carry people on the bench without work. One region was rumored to have a 2 week rolling layoff policy. Regardless of who you were, or your potential benefit to the company later, if you were on the bench for 14 days without an assignment, you were gone.

It was a tough position to be in because you couldn't book your own work. So it was actually the result of somebody else not booking work that could cause the loss of a consultant's job. Many friends were lost during that time. I hated to look through e-mail on a Friday especially at the end of a month or quarter because of the multiple good-bye e-mails. It wasn't just a PeopleSoft thing, this was happening everywhere.

Our region and our new Director Matthew was doing a great job actually finding us work and assignments to keep the group busy. There was work out there, but it became very competitive in the marketplace when customers because more concerned with the price of a consultant than the quality of the consultant.

Changes at the User Conference

There was another break in the middle of the year and our group was once again asked to present on current topics at the PeopleSoft User Conference this year in Los Angeles. Since PeopleSoft 8 was taking over our customer installed base, many of them had been asking about avoiding technology pitfalls in larger implementations or upgrades. That became the focus of my session that year.

We had a number of customer experiences to fall back on and it was easy to pull together some best-practice material as well as document some pitfalls we found in the field. When it came time for my session time slot, I had over 300 people in attendance. I was amazed at the showing, but it proved that many people were concerned about things as they moved forward with their own project.

The entire senior management from PeopleSoft were hard to find at the conference that year, setting up separate executive meetings with customers and building VIP gathering areas at the events that week. On the night of the big customer event, a couple of friends and I walked into the wrong side entrance to the auditorium apparently, and security guards physically stopped us to direct us towards a different entrance.

I held up my "all access" employee badge but he just replied, "This is the VIP entrance".

Upon finally getting into the auditorium, we saw the boxed in stage off to the side where everyone was standing in a crisply pressed suit.

One of the guys said "This isn't PeopleSoft is it?".

I remembered back to pervious conferences and gatherings where Dave Duffield or Baer Tierkel could be "bumped into" on the floor talking with just about anyone who took some initiative. Those days were gone.

With the conference over I turned back to road work and found an assignment close to home for another local Detroit manufacturer. It was a small software installation but they needed some oversight on how to build the architecture for it. I did the normal consulting thing and worked directly with the project management team to help build out the initial recommendations and the project plan for continued support.

The thing that seemed the most curious to me was the hallways in the technical and engineering wing of the building, completely filled with male employees from what I found later to be India or Pakistan. I took a break one afternoon and ventured deeper through the hallways between the desks and meeting rooms, and there was clearly a majority of non-US workers in these areas of the engineering building.

During lunch, the cafeteria became a segregated room where people easily identifiable as non-middle-eastern stayed on one side and the remaining group stayed together on the other side. Nobody really talked about it while I was there so I asked what I thought was a politically correct question about how much work was outsourced at the company.

The response was "as you can see, a lot of our work is jobbed out to low cost providers".

I simply found it ironic that not only were the US label parts being manufactured out of country, but even the white collar jobs were leaving Detroit too. "It is after all", I told myself, "a world market."

It would be a path that PeopleSoft would walk down some months later, acquiring its own India based technology company. This would be done to provide us an advantage when competing with other service firms who blended in off-shore labor in their rates to keep their bids low.

Chapter Fifteen – New York Impressions

2001 gave way to 2002 without much of a fanfare. PeopleSoft would protect the revenue they were able to obtain that year and the expense restrictions continued. Any internal travel, even for a sales opportunity had to be approved by senior management. Other companies were under similar restrictions and it was clear that everyone in the technical sector had tightened their belts.

Again the beginning of the year started like the others, a calendar year meant a new fiscal year and the project budgets were beginning to open again. I was sent to several start up projects to assist the management team in getting the technology foundation laid in advance of the project. In short, I was back on the road again.

On the home front it was an interesting summer for everyone in my family. There was too much happening all at once and trying to find time to do necessary things, family things, turned out to be difficult. There was some ice damage from the winter to the exterior of the house that remained untended and I still need to get that damaged sheathing replaced.

We never really planned a vacation and the kids started asking what we were going to do. Where should we go and when? How much do we need to worry about finding our son a new hockey team this fall? Did our oldest daughter want to give up dance for soccer this fall? Was our youngest daughter too young for boyfriends?

The economy was still in the toilet and constant internal memos at work reminded us that we needed to work harder and smarter, save money on expenses, take mandatory vacation time, give up our previous bonus structure, and take the chance of getting terminated if we were not billing our time to a client. This followed closely by another memo that told us to cheer up and keep that creative energy flowing.

It did not seem to be a friendly work atmosphere in early 2002. I was starting to receive weekly notes again from close friends about their recent termination, often without a "thank you" for their service. It was difficult to remain positive about a job that I really did like while the company, the country and friends close to home were struggling.

My wife and still best friend was struggling with other issues, not the least of which was trying to manage 3 free-thinking children without the benefit of a full time husband at home to share the load. Prior to the Summer of 2002, the kid's grades in school began to drop and there were frequent "issues" happening at home without either parent around to monitor things. One afternoon, several prank calls were made from our phone to some angry senior citizens. One elderly lady called my wife back later that day (you just gotta love the star-69 feature), to play back the recorded message and all the four letter words in tow. Thankfully it was not the voice of our son but of an easily recognizable kid who lived in the neighborhood. The rules about having friends in the house in the absence of a parent, were about to change.

Cheryl eventually had to leave the job and the employer that she loved to be home for the children. It was a decision that she said had nothing to do with me being gone most of the time, but we both knew better. I regret that the decision to leave her career had to be made at all. Things did however start to improve at home. With her constant reminder to finish their homework and study for upcoming tests, the kids did do better in school and were all headed back to the honor roll for the last semester of the year. There were no more "incidents" on the phone, and the kids seemed to appreciate the constant presence of "mom" even if that meant something innocuous like "knowing where the band-aids were". The year-end report cards were pretty good for all three kids and Summer was finally upon us.

Back in Manhattan

My new assignment was based in New York on Wall Street where the closest full service hotel was about 7 long blocks away at the Marriott Financial Center. The hotel was one of the few buildings still standing next to the former World Trade Center.

The first day was a blur after my arrival at the customer site, and the day drew to a close without much time to breath or eat or acclimate yourself to the surroundings. Somewhere around 7:30 PM, a small group of us (non residents) made our way down Wall Street and towards the NYSE. We walked slow which allowed the opportunity for a first day's impression. This was the first time I had been in New York since 9/11.

The streets were still under construction even here, with thick metal plates covering much of the road and makeshift fences blocking sections of the sidewalk. Entrances to certain subway lines were still closed and there was dirt and bits of stone and concrete everywhere. Close to the NYSE, more fences were erected some 50 yards around the building and you had to show credentials to armed police to gain entrance to the street leading to the exchange.

The 4-5 subway line was open across from Trinity Church but the streets were a construction nightmare on Broadway too. We walked past the graveyard next to Trinity, and continued down Rector Street towards our hotel. Alexander Hamilton, who was laid to rest at Trinity Church, could not possibly have imagined the devastation that existed only two blocks away.

"War zone" was perhaps the best description of the surrounding street scene. Many buildings were damaged and boarded up and permanently closed for business. There simply weren't enough people down here anymore to support the little Deli, or the barber or the souvenir shop. Dirt and rubble could still be found in the nooks and crannies between the sidewalk and the architectural façade of each building.

Further down the street you could see the light towers on the perimeter of the construction site shining down into the hole in the ground as if illuminating a sporting event. Demolition of the remaining steel and ruptured concrete footings continued day and night.

We had finally reached the Marriott hotel as it sat directly behind one office tower completely draped in black mesh to keep out birds and debris from its otherwise destroyed shell. No one wanted to go out for dinner after we checked in. Curtains in the room were closed to keep out the stadium lighting and to block the memory of what had happened here. It would be room service and perhaps a movie on cable to help remove the stress from the day and detach us from reality, if only for the evening.

Three days had elapsed before I noticed that the streets and sidewalks were wet in the evening around our hotel even though it had not rained. It was the fourth day when I spotted a water truck spraying down the entire street, washing away the dust into storm drains. The few remaining shop owners were also hosing down the sidewalk in front of their doors.

"Dust from the construction site" I first thought.

"Ashes from the grave site" turned out to be more accurate. The WTC site continued to be a devastating scene now six months after 9/11.

Spring gave way to summer and the walk back and forth to the hotel became more enjoyable. The street vendors came out in force with $2 "I Love NY" tee shirts and fake Oakley sunglasses and just about anything else you could imagine. The WTC site was mostly reclaimed now, but the tourists still flocked to the site for a view of the hole, and to remember what could actually happen in this country.

The construction workers had erected a cross out of steel beams and it stood on the edge of the site next to Broadway. Another reminder that this was a mass grave site. I too needed to grieve over the WTC events and the staggering loss of life even though I was a complete outsider. I can't imagine what New Yorkers must have gone through having lost relatives and friends there. It remained a scar on the face of the Manhattan skyline and yet the overwhelming majority of New Yorkers carried the burden with determination and pride.

Cheryl called me one day and asked me if it sounded like a good idea if we brought the kids to New York. Surprised, I remember thinking about how my kids might react to the size of the city and the masses of people here. We had been to Chicago and Toronto but usually on a weekend and the downtown areas were never as crowded as they are every day in Manhattan. "OK, Why not".

We planned a visit over a long weekend. I told my customer that I would be flying the family out on Friday morning (therefore needing the day off) and we would be tourists for the weekend. It was amazing how fast I got recommendations on where to go and what to see and what restaurants to visit. Everyone here in New York loved showcasing their own city. It was easy to fill each day with activities and although you cannot do New York in a weekend, we found plenty of "kid" stuff to do over three days. It also seemed to be, pardon the expression, the "American thing to do".

Being from out of state there wasn't a whole lot of direct support I could give the folks from New York after mid September of last year. I made a cash donation to the Red Cross as well as the more liquid form of donation, but now almost a year later this felt like the right thing to do. Visit New York, spend time here, spend money here, show the kids what happened and where it happened and teach them that Americans don't take this sort of crap lying down.

New Yorkers and the rest of the country were healing and rebuilding and fighting back. In our own small way, we wanted to be typical tourists and help bring back a little normalcy to the dynamic that was and still is New York. In short, we spent money.

I surprised my kids but mostly my daughter Jennifer when I greeted them at the airport in a black stretch limo. Jennifer had talked about being a "big shot" someday and wanting to ride around in style. They freaked out when it pulled around the corner to pick us up at the curb. The driver opened the sun-roof so we could get a picture of them all poking through the top of the car.

Every experience seemed to be their first experience. Not that we live in the boonies mind you and yes we've taken the kids to Chicago and Toronto, but they had never been to a place like Manhattan. First limo ride, first cab ride, first ride on a double-decker tour bus, first ride on a subway, and first ride on a train and those were just the transportation options.

During the weekend we had visited the Air and Space Museum, and the Statue of Liberty Island, Ellis Island, Rockefeller Center, Time Square at night, Grand Central Station, South Seaport, and the Bronx Zoo. There was also a ton of walking around mid-town and downtown, street shopping for fake watches, corner deli visits and cab rides. We essentially packed all we could into three sunny days.

The weekend sadly came to an abrupt end and on Monday morning, I packed them all into a cab and sent them back to Laguardia airport and an awaiting flight back to Detroit. Back to Wall Street for me and on with the rest of the summer.

A Clash of Cultures

That week became a personal challenge for me particularly in the area of managerial and people skills. We had a small team of technical people on Wall Street and it was a true mix of backgrounds covering both technical and cultural aspects.

Though I was the technical Program Manager on the ground at the site the team was mostly self-managing and I focused on relationship management with the customer. During one meeting which I did not attend, a mild argument broke out between a young female consultant and an older male consultant all in front of the customer. The customer had told me after the incident that it was "very unprofessional". The woman had approached me later to describe the problem from her point of view.

It seems that there might have been some cultural underpinnings to the incident, given that both parties were from different countries where women typically held lower status than men. Knowing that, it still didn't matter what the basis for the argument was when it simply violated the basic rules of consulting.

Speaking to most everyone else in that meeting first, I gathered what information I could then set up a meeting with the older gentleman involved in the incident.

Vishnu was closing in on 60, and had been with PeopleSoft about 3 years. He apparently had a distinctive educational background and had advanced to a more senior consulting position at PeopleSoft as a single on-site problem solver. When I scheduled the meeting, he did not show up. I had to look for him and ask why. In short he was angry at me for asking about the incident, but not coming to him first.

"The customer registered a complaint with me first, that's how I came to be involved". I replied.

"Well, you should have talked to me first as the senior ranking resource at this customer" Vishnu said.

"I have been on-site longer than anyone. I am the senior consultant here. The only reason there is an incident is because you say there is". He continued.

"Let me say this again, the customer registered a verbal complaint with me about your behavior in this meeting".

"When I scheduled time with you to talk about it, you did not show up". "How am I supposed to interpret that?" I asked him.

"You are not senior here and you are not my boss". Vishnu smiled.

"Well, I beg to differ". I replied

"I am the technical program manager on-site, responsible for the technical consulting relationship and I have been with PeopleSoft longer than you." It felt like I was talking like a pre-teen.

"Now I have to manage a situation that you and another consultant created". "There will be no more outbursts or disagreements in front of this or any customer." "And if I call for you to be at a meeting, then you need to be there."

Vishnu closed his self-proclaimed independence with, "I do not have to listen to you and I have no respect for the way you manage this customer."

I felt better about the whole incident after making some inquiries about this guy to the rest of the consulting management team. They seemed to support what I had found with his behavior in New York. For only the second time in my PeopleSoft career, I logged a formal negative evaluation about an employee for "borderline insubordination".

Not so surprisingly, his staffing manager protected him after the evaluation was filed. I was questioned about my motives, whether or not I overreacted, why I didn't take the time to speak with Vishnu on-site. Vishnu had told his manager that we had no such conversation.

"Ah, I see now". I replied. "I'll have the Director responsible for the whole customer account call you and explain things".

I did, he did, and the whole situation went away. On a side note, Vishnu was promoted after that.

September 11, 2002

I woke around 6 AM, mostly because the sun was up and was sadistically leaking through the gaps between the curtains and spilling into my room. I laid still for a bit and thought of the events of the previous day. The management team at our customer was asking if team members felt comfortable coming into the city on Wednesday the 11th.

"It would be OK to stay home" they were told, if employees felt strongly about personal risk or were somehow impacted with the significance of the anniversary. It had been a tough week so far and I felt like being home, though air travel on this day would have been the only thing I felt a little apprehensive about.

I was still staying at the Marriott hotel, situated a block from WTC site and I expected things to be a bit harried on the surrounding streets this morning. The hotel staff informed us that most streets around the hotel would be closed and that getting a cab would be next to impossible. I also knew from the night before that news crews and their trucks had been taking up positions in the city blocks that surrounded the WTC memorial. My head was still a bit foggy but I knew that I needed to get moving if I ever expected to get to work. It was 6:20.

 A letter from the hotel had been slipped under my door and it spoke of the events of the day and suggested that all hotel guests participate in 4 moments of silence during the morning hours. 1 moment for the time of each airplane impact, and 1 moment for the time of each tower collapse.

I remember taking a deep breath with a slow exhale after reading the end of that sentence. The hotel also offered us red, white and blue ribbons to wear during the day as a sign of remembrance. I showered, I changed, I wore my ribbon and I entered the street at 6:55.

Down the block I could see a mass of people gathering at the barricade in front of the memorial. The streets were closed in most directions around the site but you could still walk up to the fencing on the perimeter and look through. The crowds were starting to build and the street vendors who normally take position around Wall Street were starting to filter in and set up shop for the long day ahead.

I stared into the site and was transported back to the place I was a year ago when the news of the disaster first hit us. It was made more surreal as the entire office staff crowded around that little black and white television screen no bigger than a palm pilot. Once it was clear that our country was under attack, I remember my first emotion was one of rage. Now a year later standing at the foot of the WTC site, I was again filled with emotion but it was more of a subdued sadness than pure anger. I felt the need to look away and move ahead with the rest of my day.

On the walk back I was struck by a few street images that made me pause and take notice. Not that any particular street scene was in and of itself a significant event, just that they all seemed to represent a particular important dynamic of the day.

On the walk back towards Rector Street, I passed two reporters who were without cameras or microphones for that matter but they continued on with their reports over cell phones. They were facing each other and holding cell phones upside down so as to speak into the bottom like a microphone. They pointed to each other when one wanted the other to take over the report.

Rector Street was closed and free of vehicles except for a pumper-truck from FDNY engine company 204. Three young men in dark blue pants and not-too-clean FDNY tee shirts stood in front of the truck while another older fireman in full dress uniform took their picture. None of the men wanted to look at the camera, none of them were smiling.

I tried to put myself in their shoes, imagining what they must be thinking today and right at this moment. It was not a day to be happy about or to preserve with a picture of the living. It was clear that these surviving firemen did not want to have their picture taken. For them I imagined that it would be a day to honor the dead, their friends, their brothers. One raised his hand to wipe his face just under his left eye.

A block later after passing store fronts that were still boarded up, stood a representation of how some retail industries could flourish even in these tough times. At the corner of Rector and Carlisle you could still see the mixture of neon and flashing lights of the

"Thunder Lingerie" aglow even after sunrise. You can't miss it. It sits next to the "Pussycat Lounge". Grab on to a little bumpin' and grindin' to take your mind off the worries of the day, or of course, pick up a little latex novelty on your way home from the office.

As I passed Trinity Church I thought that it might be more appropriate for the reporters and street vendors and general opportunists to spend a little time here instead of hanging on the fence a few blocks down. The front of the church was void of any patrons.

Perhaps the best image of the day came as I approached South Street near the water. The corner of Wall and South happens to be next to Pier 11 and a flurry of ferry traffic from New Jersey and other places still yet to be determined. Each boat unloaded hundreds of people on the street and they all flooded lower Manhattan in droves this morning.

It was a typical morning really as the men and women carried their brief cases and coffee and newspaper while walking at a typical New York pace. What made the image stand out really, was the occasional red, white and blue ribbons attached to the lapel, or coat, or purse strap. It was a way of saying, "You bastards didn't stop us, life moves on" or at least that's what I was thinking.

We were all here to work, keep our respective companies moving forward and doing business, and we were here to remember. During the day, we shared a moment of silence on each floor of the building and everyone participated. Some were in tears. It was a tough day but one that needed to be remembered.

End of Summer – September 23, 2002

The summer wound to a close last night, with the passing of the Equinox. My time at this client was nearly over and once again I was forced to say good bye to some new friends I had made here. I have a love-hate relationship with this town. I love coming here and enjoy the fast paced nature of just about everything. I love being able to find any type of food I could want, and I especially like the people. Tough on the outside, tender on the inside. Hmmm.

Sounds like a commercial. Hate? I hated living in the city. To much noise, to cramped in the apartment, too many buildings and no trees. Simply too much for a guy who likes the slow pace from time to time. There is no "slow" in this town. Fun, yes. Slow and relaxed? Not a chance.

The Fall months were spent working on proposals with our friendly local Solution Delivery directors. We had spent time at many customers trying to put our best foot forward and win more consulting business but we seemed to be losing more contracts than we were winning. Customers became more devious as well, making verbal promises and statements about our chances for winning a particular bid, then asking for free services or demo products just before awarding the contract to a competitor.

One larger repeat customer of ours had a habit of ignoring our account management people partly because they reorganized their management structure about as frequently as we did ours. When they did call, it was always because they wanted our top consulting people to staff some internal "Quality Assurance" project but of course at no cost. They had a favorite line to be used in just such a request... "In the spirit of partnership, we feel you should be part of this effort". You need to look for the definition of "Spirit of Partnership" in the dictionary between the words "freckle" and "freedom". Ah, there it is... "free".

Sometimes we would be asked by our partners to help staff a project with key knowledge-base resources, other times our partners/ competitors would take the whole pie and only call on us when the "product" started to cause trouble. In some cases we would simply lose the bid to a lower cost estimate. Once again, its all about the money and not about the quality.

For customers this was a smart move because they could still hold the contract winner responsible if they screwed things up. Things had to be fixed for free at that point, so the low cost bid, regardless of initial quality, seemed to be the way they wanted to go.

In order to compete at all in the consulting marketplace, we chose to go build a blended on-shore / off-shore solution for our customers. PeopleSoft acquired an India based company to help with certain technical tasks on a project and allow our bids to become more competitive. It was the way business needed to be done in order to stay in business at all.

It was a mixed message to customers who still expected premium specialty resources at the reduced blended rates. Despite book value rates for some of our senior specialty resources reaching over $250 per hour, many customers were looking for their services for less than $100 per hour. It was insane, but it was reflective of the market in 2002.

One of my local buddies and Solution Delivery directors in the Midwest was faced with a more direct customer challenge. After one particular customer VP had sent our sales team "packing" with their first proposal for assistance on a project, they were told by this customer that they better come in under $130 an hour for all resources if we want to compete with the other proposals on the table.

Back to the office the team went to re-work the staffing and the "dollars" for the proposal. The only way they could get the blended rate down for the whole project was to factor in the cheaper off-shore resources. The entire weekend was spent re-drafting the numbers and the project plans to meet the latest customer demand where it was presented the following week.

The customer VP seemed pleased with the new bid but then blew up when he saw the mention of off-shore resources in the statement of services. He continued to berate our Director on her lack of "Patriotism" by not using "only" American labor in this proposal.

She defended her position by telling him that they had "forced" us to do that based on the "ceiling" price of $130 per hour "that you mentioned in our last meeting". The customer VP claimed that no other proposal he received had mentioned off-shore labor in the bid, and we left by advising him that he should check on that before making any further assumptions.

Maybe it was embarrassment, maybe it was another reason but the customer chose to go to a different discounted vendor for the work. He later called to apologize to our director and did indicate that the winning bid contained off-shore labor. His management team however, forced him to take the cheaper path regardless of the potential loss of quality. To me, its all about budgeting properly, qualifying project risk and managing internal expectations. If you can't do that, you shouldn't be making multi million dollar decisions on behalf of your company. Someone will always want you to take the cheaper bid.

The PeopleSoft Consulting organization continued to change. In each of the years 2001, 2002, and 2003, the total compensation package for consultants and project managers was re-organized to "bring us in line with industry and competitive marketplace standards". You may not need a college degree to figure out that it potentially meant a reduction in pay. This also seemed to be a common practice with other firms in the industry as well. It was just another sign of the economy in stagnation and companies doing everything they could to reduce operating costs.

Our Consulting Services VP also continued to build up the "Practice Initiatives" organization which started to take ownership of customer issue escalation responsibilities as well as the management of certain knowledge capital. Consulting "knowledge capital" including something called "packaged services" which included a number of fixed-length service products that were developed originally in the Midwest.

There was some resentment in the region over the P.I. group just taking ownership over something we had developed, but it was not a corporate climate that tolerated negative upward feedback very well. Stephan knew that all to well.

Focusing on Important things

It had been a long travel year for me in 2002 and much time was spent away from home. Because of the dates of my daughter's respective birthdays falling on weekdays this year, I had been out of town and had to settle for a phone call at the end of a typically long day. On weekends, I usually spent time with my son in hockey rinks traveling from city to city and state to state in one travel tournament or another.

Cheryl told me late that year that the girls made comments about "dad never spending any time with us". It was painfully true. They were "big" into dance classes which were held during the week, and hockey happened primarily on weekends. When home, I wasn't much for playing with Barbie's or changing hair styles like they did in the privacy of their shared bedroom. It wasn't done on purpose, but I did wind up spending more time with my son.

On a whim, I had called another corporate buddy who lives in Portland Oregon and struck up a long conversation with him about things to do in the Pacific Northwest in November. Kenny loaded me down with tons of ideas and web site links for national parks and forests and shorelines. I was able to map out a long weekend trip between Washington and Portland.

"But why here when its almost Winter?" Kenny asked.

"Because I've never been to that part of the country, and I wanted to take my daughters on a photo weekend."

"Ahh Bonding" Kenny followed.

Both my daughters had taken up a fondness for photography during our last two family vacations. After I would pull out my camera to capture an image, they would ask why, what, how. We started them on disposables first, but then bought them both a moderate priced 35MM auto focus camera. When I told them about the trip to the West Coast, they flipped and started packing about 3 weeks in advance of the flight "just to be ready in time".

"Daddy, will I need a coat?".

"Yes, but you better pack 3, just like your mother."

Kenny and Tina had offered up their home for our lodging upon arrival to Portland. PeopleSoft people were like that. "Need a place to stay?" "Take my bed, I'll sleep on the floor". "Make yourself at home". "Does that mean I can walk around your house in my underwear and stand in front of the refrigerator with the door open?" I asked.

The girls stayed awake during the entire flight which was amazing to me, since they both drop limp about 5 minutes in the car after hitting the road. Maybe it was adrenaline, maybe it was the endless supply of cola and caffeine from our flight attendant who thought they were "so cute". We eventually arrived at the PDX airport and found our rental car without trouble. Kenny and Tina were about 15 minutes away.

Their turn of the century home was about to be placed on the national register for some architectural characteristics that made the home unique. It was beautiful and adorned inside with most of the original wood trim, floors and cabinetry. The girls gravitated to Kenny and Tina's new baby who was making fast time on the floor between plastic toys. Suddenly all three children were entertained.

The next morning we all headed out to discover a little history in the surrounding Vancouver Washington area, and despite the drizzle, the girls found camera shots that they could take just about everywhere. Later in the day we said goodbye to our gracious hosts and headed up the west coastal highway ultimately to the top of Washington State in Port Angeles.

We stopped at one of the few actual rain forests in North America along the Washington coast, well north on highway 101. It wasn't raining at the time but as we hiked inward on the trail from the parking lot, the tree canopy was still dripping on us fairly strong. The girls would not be deterred however.

My youngest found a bright, almost fluorescent yellow banana slug on a moss covered tree root which made for a good shot, and my oldest daughter found a light green very small tree frog on a wide leaf. They shared shots of leaves and flowers that we've never seen before and it was fun to watch them get excited about photography.

We finished our second day by finding our hotel in Port Angeles near the bridge to Canada. The next day it was all about the northern shoreline and sea lions. Marine wildlife was everywhere and even in the distance, too far for our camera lenses, you could see large fin activity on the horizon. Kenny had mentioned that some of the whale pods stayed around the north waters a little longer than others but we were never close enough to capture a shot. Still the girls and I found some great detail in the boats at the marina, waves, sea lions, and an occasional pelican or two. We finished out the day with a movie and a pizza and then they both bounced off the walls of the hotel jumping from bed to bed and coaxing me into a pillow fight. <ahem> I won.

The last day of our trip took us around the eastern side of Washington as we headed to the Seattle-Tacoma airport. It was cloudy while driving up into the mountains but after our ultimate decent, you could see the cloud layer level off just above the window line of our car.

"Girls... open your window and touch a cloud" I blurted. And they could.

I slowed the car but the cloud stayed absolutely level, just above our window line. Kinda strange, kinda cool. A fitting image to end our special trip. The girls talked all the way home and made me promise to take their pictures to the one hour photo place.

I told them, "Tomorrow, we'll be getting back late tonight", then I tried to explain the time zone impact of flying west to east.

"Something I knew all to well" I thought. Finally after a long flight back, just as predictably, they were asleep 5 minutes after they hit the car for the ride home.

Chapter Sixteen – Six Horrible Days

Early in 2003, I was able to spend time at home again thanks to two separate efforts going on with local companies.. It was just a checkpoint audit for two existing projects and pretty low on the difficulty scale. Still, even with the sometimes long days it was good to get back home, rest in my favorite spongy chair and tease the kids about their school day.

The company was still doing pretty well despite a tough economic climate. Even though Bush was claiming that the country was on an economic rebound, the whole of the technology industry was not seeing the benefit. Overall project work was down and software sales were down.

The only way PeopleSoft could grow was by releasing new products, taking business away from their software competitors and reducing operating costs. Quarter over quarter, we seemed to be able to run the "cost-containment" ship pretty well, but that included things like cuts in training, cuts in technology to support field operations, cuts in our bonus payouts and cuts in travel for just about any kind of internal business.

At the end of January, I finally collected all of my tax information and started to go over the particulars of my W-2. It turned out to be the third consecutive year of reductions in total compensation. That along with the sting of watching the value of retirement funds wash away through 2000, 2001 and 2002. We could do nothing but sit and watch it happen.

The country seemed to not only be struggling with the loss of manufacturing jobs, but also white collar technology jobs as well. If you were in the market to hire computer aided drafting labor, mechanical engineers, or computer programmers, you were now looking off-shore as an option for cheaper labor.

News reports on statistics released by the US labor department highlighted statements like "New requests for unemployment benefits were down in the fourth quarter". As if that told the whole story.

The headline never includes anything like "The millions of people whose unemployment benefits ran out, are still unemployed." I talked frequently to two close friends who were both in the Information Technology field. One has been out of work for more than a year, the other had to uproot his family and move out of state for significantly reduced pay. Both of these people I would classify as brilliant, dedicated, and experienced but those qualities were not enough to get hired in 2002 or 2003.

Spring led to summer and our family was spending as much time as we could together on weekends and during the week between the girls dance classes and our son's hockey. There were no glamorous vacation plans this year partly because of the cost concerns, but also because there was too much going on with the kids activities and house maintenance. We would plan long weekend trips locally or spend a week in the mid part of Michigan where the family had a cabin. All in all it was a relaxing beginning to the Summer.

The Technology Solution Delivery Group was all but a memory by the time 2003 began. The Midwest was the only region to keep its group together and the regional management team had to defend the benefit of our group to Executive Management in Pleasanton. The other regions had lost most all their senior global architecture people through attrition, termination or absorption by other project management teams. The focus of our group was supposed to be, according to the consulting group leadership, "project management" not "technology expertise".

The Practice Initiatives Group continued to take a leadership position for customer technology, architecture, and tuning issues. If customers experienced problems in the field, this new group stepped in for the issue escalation process. This was a role we had built up in the field in our Solution Delivery group, but was seldom used.

I felt that part of the trouble with some Pleasanton based technical groups was that they were completely reactionary. Yes the corporate group was effective in defining new standards of operations, new techniques, and new tools. But it would only apply those new tools and techniques when a problem had been identified by a customer.

We seemed to lose sight of the fact that customer issues could be avoided if we could help guide them past the pitfalls, plan the appropriate architecture, build the support infrastructure, and train them how to manage a complex integrated system like PeopleSoft. That is the benefit our group provided in the field and yes, very successfully. Proactive versus reactive. Pay me now, or pay more later. Not to pick up someone's brick and throw it back at them… but since we're supposed to be project managers, "planning" and "risk mitigation" are part of project management "best practice".

Our struggling group was not being marketed or supported in the field as a necessary component of the sales or project life cycle. Often in a competitive bid situation, the technology oversight hours were cut out of the proposal in order to reduce the bid. Many proposals for project work were delivered with little or no field experienced technology consulting. Our team in those situations, was only involved with the customer only after customer satisfaction with architectural reliability or system performance, was negatively impacted.

Our regional group hung on, and we were often called upon by other regions to help them out with projects or quality assurance site visits. Though we were never supposed to leave the Midwest, I was called to help customers in Portland, New York, Dallas, and New Jersey in 2003. But no, our little regional proactive model didn't work. Sarcasm? Nah.

The 2003 User Conference was upon us again and the Midwest Tech Solution Delivery group led all regions in approvals for technology sessions at the conference. I focused on technology project risks again, preaching the "technology planning" aspect to a project's success. We all prepared our sessions in advance and started looking at the attendance numbers for those customers who had signed up early. It looked like another large crowd this year and once again prepared myself to conquer the fear of being on stage.

2 Days in June

A flurry of e-mails started to hit our in boxes the first week of June announcing PeopleSoft's plans to merge with J.D. Edwards. The offices were buzzing about what the ramifications might be, including the possibility of job cuts as a result of the merger. More importantly, the merger would combine two distinctly different product lines, for a mostly unique customer base and pushing us past Oracle as the number "2" enterprise software vendor in the world.

The synergies were vast and the cultures of the two companies seemed to fit very well. Everything we were reading about the proposed merger was good news rather than bad. Now I realize that much of that internal news was designed to make it all look wonderful and cooperative. Happy, happy, joy, joy. But even Wall Street liked the merger and interviews with the JDE senior management team indicated that they were excited about the news as well. There was plenty to talk about in the offices that day and smiles seemed to be everywhere.

On that Friday, Oracle announced a hostile takeover of PeopleSoft followed by a few well placed comments about what would likely happen to the employees, the support of the current product lines, and the technology. Much of what was said in the press was "spun" in the weeks and months that followed as each company was trying to paint themselves in a better light.

Customers were clearly upset and asking questions about the proposed takeover wherever I went. It felt like I had to engage in damage control first, then get down to business if any business could be done at all. It was difficult for me personally to make progress at some customer sites, because they became apprehensive about budgeting for any new upgrade or product install until the whole Oracle thing was resolved.

Some of our larger private sector customers and government customers as well had just made huge investment in the PeopleSoft infrastructure prior to Y2K and after, to be the planned backbone software system for the next 5 to 10 years. To change software or platforms now was a capital budget expenditure nobody was planning on.

My wife, my father, my in-laws, aunts and uncles and friends all wanted my take on what it all meant. Was my job secure? What was going to happen to the company? When would I know the decision? When would it all be over? I think I realized then that I was lucky just to have people concerned about me and my immediate family. The short answer was though, that we had no idea of what was to come and that was perhaps the most terrifying thing of all. Not to jump too far ahead in the story but after months and months of adjusting the offer per share, extending the offer to buy, declaring their intent to see the takeover through to completion, I felt bludgeoned.

More Cost Containment

The offer put more pressure on PeopleSoft's ability to control costs. One of the measures was an effort to eliminate the travel company that we had been using since I arrived at PeopleSoft. There was an internal push to use a web-based automated system for booking flights and hotels rather than speak to a real person. OK, I can see the cost savings argument if we weren't passing the cost on to the customer as part of the "reasonable and customary" expense policy. Ticket bookings always carry a booking fee even if the front end reservation system is a computer. I took this new policy personally, because my buddy Francesca had been watching my back on travel since 1995. One phone call, 5 minutes and I could get everything booked. Not 20 minutes trying to figure out strange airport codes and the closest hotel.

There was a definite difference of time spent by the consultant, whose opportunity cost was between $200 and $275 per hour. The agent reservation system worked well, because they knew who you were, and they knew your preferences. They researched the airport codes and the distance from the customer site to the hotel, and what hotels had poor ratings or were in questionable parts of town. All things the automated system couldn't provide. And if you needed to make a change after booking, you talked to the same person for correction not a new guy after waiting on hold for the next available operator.

You could make the argument that the automated system was just as easy, if you were traveling to the same destination, knew all your flight codes, knew where the hotel was and the name of it, distance to the customer site, rental car availability versus public transportation...... yes it could be easy. Not if you were all over the place, each and every week.

Francesca rescued me from many more than one bad flight and many more than one bad hotel. That level of service was something that needed to be paid for even if it was a personal expense. The days of travel agents are now gone from PeopleSoft and I still can't figure out where the cost savings came in.

2 Days in September

The show must go on. Back in New York again after a year's hiatus, our group was called again to lead a global architecture build effort for one of our largest distribution customers. Again the region in question had nobody left in Technology Solution Delivery to staff the demand. I happily took the 5 week job and settled again in Mid-Town near Time Square. The analysis phase of our effort was running along pretty smoothly through September and then I received a tearful phone call from home.

One of my remaining two uncles on my father's side died suddenly after what seemed to be a full recovery from a hospital stay related to his Diabetes. My father was distraught but seemed to be holding things together as the oldest brother in the family, but my wife indicated that he seemed "a little shakey". This was a major blow to the extended family.

My uncle Randy was the father of 5 children and a bus load of grandchildren and nobody was prepared for his passing. My dad and I talked that night again, but the funeral arrangements hadn't been made. It was likely that services wouldn't be until that following Monday.

I was worried about my father, because his other brother was also in the hospital and Dad was still dealing with his own loss of energy and mobility after a cancer fight himself. I made plans to return home a day early from New York and thought about family all night. There was no sleep. Its this kind of event that forces you to look at things and evaluate what is important. What would happen if something happened to my wife or kids and I wasn't home? Worse, out of state, or out of country.

The next day the second bomb dropped. My only other surviving uncle, my Dad's youngest brother, gave into cancer early that morning. Two uncles, two days. Two brothers lost for my father. I can't imagine what he was going through then, much less the extended family who was still trying to deal with the loss the day before.

The funerals were held on two consecutive Mondays to space the grief apart by at least a week. All the family was there, and it was heart wrenching to sit behind each family and see the grandchildren of each proud man emotionally affected the most. I had to look down at the floor after a while, because I just couldn't watch them anymore.

I wondered again if I was still doing the right thing. Was our family still functional, was it all still working, was I just pursuing a job that I loved without regard for the needs of everyone back home. I asked a thousand questions of myself during those two weeks but I still had no perfect answer. Cheryl was supportive as always, and she of course handled everything at home whether I was there or not. I still needed to be happy at home and at work. She remembered the days when I worked for the insurance company, 7 days a week on-call. The management team was actually verbally abusive to its employees. "Feel lucky you have a job" was stated on more than one occasion. "You need to work whatever hours are needed over the holidays to get this thing running before year end, my bonus depends on it". No shit, I'm not making this up. That was an actual quote.

Everything was a little foggy for the next few weeks and I returned to New York to finish up on the documentation we needed in order to satisfy the contract. The work continued in other states and I split my time between the home office and the road as we headed toward the holidays.

2 Days in December

My parents live relatively close to my own family but after years of training me to be independent because of the way they practiced their own independence, it wasn't a situation where we saw each other with much frequency. Of course there were visits and "stop bys" and planned dinners or holidays, but it had been a few weeks since I had seen them. I decided to call one evening after returning from wherever I was that week and my mother mentioned that Dad seemed to be constantly tired for some reason and wasn't eating. I asked a few questions about his growing pile of medication and wondered if he was mixing the AM/PM meds.

My wife used to be a medical assistant and was pretty good at recognizing the names of most medications and their intended use. We grabbed her medical reference guide and headed over that evening. Dad was sleeping in a chair when we got there and couldn't stay awake long enough to even carry a conversation. I asked how he felt, which was "fine just tired" he said.

We found on his dresser over 20 different medications, some expired and no logic as to what was being taken when. He couldn't explain what was being taken either and my mother had no idea if he was skipping some or taking them properly. He kept repeating "I'm fine, I'm taking what I'm supposed to". We wrote all the medication names down and went immediately to the pharmacy to break down the combinations, and the side effects.

There were some that cause drowsiness, and others that couldn't be taken in combination. Some were old and shouldn't ever be taken. It was a mess. We documented everything and bought a 7 day pill divider to separate the morning and evening drugs then headed back to my parents house. I asked again how he felt and he got annoyed that I kept pushing the issue. Cheryl made up the pill box combinations for 7 days and we left around 9 PM with the plans to check in the morning to see of the drug combinations helped at all.

The next day, the news from my parents house indicated no improvement and my mother made the comment that his hiccups were getting worse. I asked if she had called his doctor, but she wasn't sure who the primary would be based on the specialty doctors he had been seeing. Cheryl had called their house as well in the afternoon but again no apparent change in his condition.

I came home early that day from work and we made another visit to see how bad things had gotten. He looked even more pale and had trouble supporting his weight, and his hiccups continued non stop. I wondered why it had to be me that made the decision to get him to the hospital since I didn't actually live there, but Cheryl and I carried him out to the truck for transport.

I took him into emergency while my mother did all the mandatory insurance stuff. The doctor immediately started checking vitals and tried to start a drip but couldn't even raise a vein. Dad had become so dehydrated that the Doctor started yelling at me for letting him degrade to this condition.

"Wait a minute", I said. I don't live with him, but I'm the one who found him like this and brought him here".

He was rushed upstairs into a corner room of the ICU and only then did it start to sink in just how bad things were. The attending Doctor finally arrived and looked at the blood work, then pulled me out into the hallway.

"His kidneys are almost completely shut down and he is very dehydrated." "With kidney failure, sleepiness and loss of strength are part of the symptoms, so were the hiccups."

The litany of ailments was outlined for me and even emergency surgery was out of the question until or unless he stabilized. My mother was a wreck and more angry than anything else for some reason, the doctors never seem to explain everything.

Nobody said the words, but I found out the next day from his regular doctor that they had only given him a 25% chance of surviving the night. I had stayed that night and listened to him struggle to breath while they woke him every hour for injections that were designed to cancel out the poisons that had built up in his system. He was just as bad off the next day, but the treatments seemed to keep the fluids running through his system. The first round of dialysis was scheduled and he remained under close watch in ICU for 3 weeks, right through the holidays.

On day three of the ordeal, I returned to work completely numb. I remember trying to go through my backlog of e-mail and eventually got out to the internet for any news on the Oracle thing. Needing a shred of hope about something, the first article I ran across was Oracle's announcement to start a proxy fight for seats on the PeopleSoft Board of Directors. I remember seeing another good bye e-mail from another buddy, after 7 years with the company.

December sucked for many reasons. Even my father in law was admitted to the hospital to have a kidney removed. It seems that a growth in one kidney has taken over to the point where it was completely shut down. It had been there for years apparently but only found recently. Again another strain for Cheryl, the medical one in the family who had to interpret all the doctor-babble into terms her own large family could understand. For two weeks, we alternated trips to one hospital in Macomb county and another in Downtown Detroit.

With my father and father in law both fighting cancer, it seemed to be a topical discussion with my mother one evening at the hospital while my dad slept. I learned from my mother just how many of my immediate relatives had either battled or died from cancer in the last few generations. I was not happy to find out just how hereditary this would be.

My boss Matthew was sympathetic to everything going on, but he was one of the few that knew about my father and father-in-law. It was not a topic I could keep talking about then, because the news kept rolling in about surgery, and dialysis, and additional cancer found to be the root cause of the trouble. The health problems would continue for both "fathers" months into 2004, but eventually they both came home and continued the fight with some regained strength. The health issues would remain a constant problem for them both from now on and I made an immediate appointment for myself to get checked that same month.

Though I didn't live at home with my parents, I would still be called on to talk about finance trouble, or problems with the car, or more medical follow up. I found myself managing some of my parents issues along with my own. It wasn't a role I was ready to take on, being a part time husband and father as it was. But there it was, it had become my time to have to deal with these things.

The blur continued well into 2004 and none of the news from Oracle or Wall Street was positive for PeopleSoft. It was as if our epitaph was already being written by analysts appearing on CNBC, or CNN or anyone else needing a headline to fill the programming void for the day. Oracle at the time could only acquire between 6% and 7% of the outstanding shares, and never seemed to convince the "average" stockholder that their offer was really the best deal going. Still the offer renewals kept coming, and the news reports kept rolling stating that it was "only a matter of time".

I tried to keep my head focused on my own wife and kids and the work in the region. Sometimes work is the best remedy for the head, when you need to avoid thinking about other things. Not that there aren't important things to think about, but they can dominate your day if you let them.

Chapter Seventeen – A New Global Customer

Out of region again, I was asked to travel to Ft. Worth to take up a small team of technical people on the ground trying to trouble shoot some performance issues on a DB2 financial solution. What started as a short term site visit grew into a multi month effort where we looked into re-designing the entire architecture. This request from the customer came after a much larger team of consultants from a competing firm failed to gather the functional requirements and translate them into a technical strategy.

They had apparently been there for a year with 15+ consultants, we had a couple of months and 6 people to figure it out. The complete redesign was also planned to account for a number of new PeopleSoft products and users the customer would put on the current systems.

Current systems + new load = New systems. I love keeping things simple.

We worked more than we played in Ft. Worth, covering long days and not doing a whole lot after work. We did do a couple of team dinners, but for the most part everyone was just whipped after the work day. Must be a Texas thing… "Rode hard and put up wet".

We managed to catch a couple of minor league baseball games at a local park put up by Dr. Pepper for a team called the RoughRiders.

Good baseball, cheap tickets, great food. When a home team player hit a home run, the ushers would walk around with a cowboy boot to collect money. Half would go the team's charity and the other half would go to the player. It seems in the Texas League, players didn't make a whole lot of scratch. It was a cool custom.

This was a challenging project environment with many individual silos of shared service employees, none of which were full time to any project. The management team was challenging to work with as well, valuing a fixed weekly schedule of meetings in favor of open time in the schedule to get the actual work done.

The customer preferred and often called meetings to discuss topics, but with no agenda and no decisions made in the meeting. Each meeting was an open discussion forum with virtually no leadership. I remember one quote from a project manager there during a disagreement about which calendar software to use for scheduling and tracking meetings. "I think we need to schedule a meeting to discuss how to schedule meetings", and he was dead serious.

For those meetings we controlled, we published an agenda, expected attendance list, advance questions and reading materials, logistics for the meeting and toll free conference number. It was actually efficient. What a concept.

The Longest Week of my Career – ended in Texas

One of the last weeks with this particular customer turned out to be one of the longest for me since joining PeopleSoft. It started with a sleepless Sunday night, but it always seems to be like this before an early flight out the next morning. This week was supposed to be easier than the last 8 and there wasn't really anything planned for the week that should have elevated my stress level. Still, "Here we are again on familiar ground" I thought.

I looked towards the glowing red digital image on the night stand, closed my eyes again and rolled my face into the pillow muttering "Fuck" as the word was muffled by feathers. It was 3:00 AM. Agitated and constantly tossing in the sheets, I tried to force the onset of sleep knowing the small battery-operated alarm in the bathroom would awaken me at 4:00 AM in time for the long drive to the airport.

I watched the last digit of red numbers transition from 4 to 5, then from 5 to 6. This alarm was set for Cheryl at 6:00 AM, so that she could get the kids off to school. I thought about how much I was willing to pay for that extra two hours of sleep and also for not having to fly anywhere this day. 9 to 10, 10 to 11, the minutes counted on and Eyes closed again hoping for a few minutes of real sleep before the smaller alarm on the bathroom counter went off.

Thoughts of the weekend crept into my head where after another 90+ hour week on the road put me in a position to forget my wedding anniversary. It was logged in my Pilot and I had even thought about picking the right restaurant, but the enormity of the previous week was overwhelming. Meetings and travel schedules were so bad that week, I never opened my Pilot to check dates.

She was understandably angry, but wouldn't even allow me to buy her some jewelry while we were at the mall that same weekend. I turned over again and flipped the pillow over to the cool side.

4 AM and the digital noise started creeping past the closed bathroom door and into the bedroom, enough so that I could hear it but not enough to wake my wife. I dressed and did my hair in silence, turning the lights off in the bathroom as I entered the dark bedroom to grab my suitcase, efficiently packed the night before. I went down the stairs but then stopped when I heard a faint jingling noise.

The family dog appeared from the girl's bedroom and trotted towards the first landing of the staircase. "Jake" buried his face into the back of my leg, hoping for another scratch. Jake got his scratch behind the ears and I made my way towards the door while Jake turned back towards the girl's bedroom and a still-warm comforter laying on the floor waiting his return.

Airport Security

Maybe it was an omen that I didn't recognize at the time, maybe just fate, maybe it was the fact that my tickets were issued at the very last minute over the weekend for this flight, but "sleepy me" was flagged for security search. The contents of my briefcase were dumped out for thorough inspection, all the laptop cords, flash memory drives, pens, pieces and parts were taken from their neatly stored little places and spread all over the table. The security agent then tried to dump it all back in, unsuccessfully.

Flight # 2173 was held at the gate because of Chicago O'Hare inbound traffic, and I sat with my eyes closed and my 37 inch inseam legs stretched out and crossed at the ankles. I was luckily able to change my seat to an exit row window, just for the extra 6 inches of leg room. Ultimately the flight took off and was only delayed 45 minutes. While I rested, I did not sleep but spent my time thinking about all the document changes that needed to be made this week in order to satisfy contract agreements with this customer. "It will be a hellish week".

Arriving onsite at the company offices in Westchester Illinois, I found the local convenience store and grabbed some banana walnut mini muffins and some cold carbonated caffeine. From there I found the conference room that had been set aside for our team on the ninth floor and I plugged my laptop in to begin the long day of typing.

People started to arrive from other cities with staggered flight departure times, and the conference room quickly filled with familiar people as well as new faces. Everyone had different expectations on what was to be accomplished this week, and everyone seemed to be empowered to move in their own direction. The trouble was that I was being assigned as the primary contact for different work streams, or at least the question and answer man for many of the topics being discussed.

Late in the morning, I was pulled from the conference room by one of the customer's program managers, the product director and the account executive. I took a seated position in the mini lobby outside the conference room in a comfortable chair against the back wall. The three managers pulled their chairs in a semi-circle facing me. It felt like a parole hearing. Not that I would know anything about that. I watch movies, you know.

"We need you to go to New York to help the international team with their budget numbers that are due this Friday".

"When" I replied.

"Tonight" was the answer.

I found out that the plan had already been made, without even the courtesy of checking with my boss whose office was just down the hall.

"You'll be there for two days through Wednesday, then you'll need to be in Ft. Worth on Thursday/Friday for other meetings with the program manager and the customer's CIO".

"What about our deliverables?" I asked.

"You'll need to work on them in between meetings and phone calls… you should be able to get it all done."

I had been counseled before in my annual performance reviews, that sometimes I wear my emotions too easily on my sleeve, and facial expressions often told the world just how I was feeling… This time I worked hard to keep my face in a firm position, nodding my head up and down from time to time as if to imply agreement. There was simply nothing else that I could do but comply and absorb whatever additional work was being put on my plate. "OK, I'll make the arrangements" was all that was said in reply.

The call to the travel agent went well, except for the calculation of the price tag for this last minute three-legged trip. It would start at O'Hare, go to the Westchester County Airport in White Plains NY that evening. Two days later, I would fly from Westchester County to O'Hare, and connect into Ft. Worth. A day and a half later, I would return home Friday from Dallas Fort Worth to Detroit. All for only $1600 in addition to the $1300 I paid for my original tickets and now defunct flight schedule for this week.

Making the arrangement changes was painful, as flight options were limited and no refunds were to be had on the original flights. In the end I was able to book the new itinerary and had another hour to put things in order before having to leave for O'Hare.

Last minute ticketing meant once again, a strip search. The O'Hare security detail pulled me aside for the "not so full Monty". No clothing is actually removed, but the shoes have to come off and the belt has to be opened and the inside of the pants buckle needs to be manually checked for hidden objects. Some stranger's hand does actually go inside your pants to check the snap, and that was all I really objected to albeit not verbally.

The contents of my neatly folded clothing was once again tossed inside the suitcase, and my computer bag was dumped again just for shits and giggles. "Its not too late to turn around and go home." I thought again. "I could be sick." I tried to convince himself of the lie that would never be. "Duty, honor, integrity and other military words…" I grinned and shook my head and packed my things for the third time this week.

Weather patterns were brewing in New York, but the gate will never tell passengers these things especially with pre-paid tickets. As the minutes passed by the scheduled time of departure, the sign above the gate was changed from *"ON TIME"* to *"DELAYED"*. "Really? Nooo. Go on. You don't mean that."

Minutes became hours, hours crossed into the next day and the plane actually departed after Midnight for another very long day. Arriving in Westchester at 2:30 AM Eastern time, I picked up my rumpled body and dragged my feet to the Hertz counter. I found a map, and some idea of where to find the Marriott Renaissance hotel in White Plains. I had taken the precaution of printing Mapquest directions so I chose not to ask the Hertz agent for local directions. Another mistake.

Wanting nothing but sleep, finding the hotel in the dark on roads surrounded by hills and trees and missing street signs proved overwhelming. I finally called the hotel asking for help and the friendly desk clerk told him that I never should have trusted the web maps. "Most people can't find us the first time." Sasha said. You need to double back and find the highway.

Sasha went on to give directions but I couldn't write them down without stopping on this two lane twisted road. I had to call back twice to ask for help finding the Hotel, which was buried up in the hills of another tree lined road whose sign was hidden by branches at the top of an apex in the road. I caught the name out of the corner of my eye, slammed on the brakes and backed up into the apex at risk of getting rear ended, just to find the driveway.

"Thank you Sasha" I said with a smile as I approached the front desk. "I had to cash in my male gene that isn't supposed to ask for directions, but I appreciate the help getting here." "Please help me get to a room so I can finally sleep."

It was 3:30 AM Tuesday and I couldn't sleep. My head was filled with the events of the day and the promises of arrival times to the Tarrytown NY office this same morning. "8 AM came early", I pondered, planning how early I would have get up and iron my clothes and make my way to the office.

The alarm went off at 6:00, and I wondered when I actually nodded off. I figured it had to be sometime after 4:00, but a nap was preferred to no sleep at all.

It was a nice shower here at the Renaissance Hotel, with a curved curtain bar that gave more room to turn around and wash in the tub. The hot water felt good on the back of my neck and I thought of home and the kids and the still angry wife. I shaved in the shower, dried off in 13 seconds (I had a system for drying off), and grabbed the ironing board from the closet. 15 minutes later I was dressed and ready to go.

With the sun up, I studied the complimentary map provided by Hertz which didn't go beyond White Plains on the detailed side of the

189

map. The more general map on the reverse side, showed a complex road infrastructure, but very little in the way of "grids" throughout the area. Lots of parkways and highways and exits that only let you off the highway, but no way to get off and reverse direction. I had another Mapquest map in hand and I started the supposed 15 minute drive to the Tarrytown office.

Roads were closed for construction at the dam over the reservoir, detours were confusing, highways and parkways provided no way to turn around, I was lost. Cashing in my male "direction gene" for the second time this week, I found a gas station with an actual American citizen behind the counter. The man named "Vic" smiled and offered friendly and helpful advice and looked as if he did this all the time.

Arriving late, the office staff laughed at the ordeal and agreed that you just had to figure it out once on your own, and go through the indoctrination. I was indoctrinated alright, but because of all the doubling back, was still unsure of how to get here from the hotel.

The day dragged on, but the meetings were helpful both for me and hopefully for the international team as well. Lunch was just about average, based on my experience with many customer cafeterias, and I enjoyed a few minutes in the sun as the clouds cleared away just long enough to cover the lunch hour. Despite all the hype associated with my last minute trip out to New York, the customer seemed accommodating and the topics for discussion were reasonable and anti-climactic.

I wrapped up the meetings for the day and had planned an early exit from the office. I thought about room service and a comfortable bed and watching a little cable to wrap up this very long and arguably wasted trip. I thought about some favorite lyrics to a song from a few decades ago by the Who, "This ain't no social crisis, just another tricky day for you.... Fella". Just about then, my cell phone rang. It was the global CIO for this customer calling for a meeting that started at 7:00 PM.

Feeling the weight of the day, the weight on my feet and the sticky feeling you get when working in an office on a hot day after the automatic ventilation system shuts down, I found my way to an open office and started the two and a half hour call. No one seemed friendly on this impromptu conference. It was attended by about 8 senior managers, some of whom I had never met, but everyone had critical feedback on a particular document I authored about a week before.

It was brutal at times with comments thrown over the phone line like "these drawings are just wrong".

"This conclusion is misleading".

"I thought it was a good effort, but it missed the mark". Not sweeping endorsement, but interesting comments from people who knew nothing about what I was writing about.

Thankfully someone else was acting as scribe and I needed only to sit back and take the verbal assault, albeit "professional not personal". "Sure, right, you bet", I thought. At one point in the call, I found the mute button on the phone and shouted a few Detroit-based, blue collar, four letter comments back at the people I couldn't see.

It ended at 9:45 PM, and I sat back for a few minutes waiting for the shaking in my fists to subside. I gathered up my laptop and looked for my keys and struggled to remember where I parked my tiny Buick Rendezvous rental.

Surprisingly, the hotel was easy to find going the other way. None of the roads were closed and I managed to work my way through the local streets. 20 minutes later, I found the entrance to the hotel. I was shocked actually, given the amount of drive time spent finding the office the first time. It was just after10:00 and I knew I missed the regular dinner menu from room service, so a sandwich would have to do. I ordered. I ate. I showered. I laid my head to rest.

1:52 AM Wednesday. I couldn't sleep. Watching ESPN news for a while, then switching to HBO, I wasn't paying attention to anything being said on the 21 inch screen. my stomach churned thinking about my long day ahead tomorrow and my late flight out to Ft. Worth in the evening on Wednesday. The channel I didn't watch was the weather channel where a clue about my upcoming travel day would have revealed itself.

Somewhere after 3:00 AM, I did close my eyes, but when I was awakened at 6:00, it felt like I didn't sleep at all. The routine so familiar now would start again. Throw the blankets off the bed, head for the shower. Kicking the boxers off with my right foot and propelling them into the open closet with the rest of my dirty clothes. "A small satisfaction to the morning" I thought, "only guys have mastered the directional tossing of underwear with their feet". Showering, shaving, shitting, ironing and dressing, the hair was last. I had planned to hit the streets by 6:30 this morning to arrive at 7:00 AM. I checked with the concierge to validate a better path to the little borough of Tarrytown and felt a little more confident about my drive to the office.

15 minutes into the journey, I missed a turn to the left. There was no street sign of course, but the concierge knew where the street was so therefore so should everyone. I tried to turn around, but was quickly forced to merge onto a parkway and that was the end of trying to follow the map. You see, in that part of New York, there is no turning around on a highway or parkway. You can merge onto another parkway, exit from a highway into another borough or drive around and enjoy the view.

I stopped somewhere again to look at the useless Hertz map and tried to determine exactly where I was. I spotted the sun off to my right and knew at least that I was currently headed north. I pulled back into traffic and found a sign indicating that the town of Pleasantville, was the next exit. Pulling over again to look at the map I found "Pleasantville" to the north of Tarrytown. There was a gray line on the map with no marker that appeared to lead from Pleasantville straight to Tarrytown so I tried to find the road in the center of town. It was 7:15.

Twisting around the narrow two lane road heading into Pleasantville, I noticed a hunched figure in the center of the street and traffic stopped in the other direction. I slowed to see a weakened old man trying desperately to hold up a stop sign and allow children to cross the street to a school on the left. I felt the scene fade to black and white for some reason. I noticed three girls approaching the makeshift intersection skipping together in unison. I spotted a stay-at-home dad pushing a stroller and walking with two older children towards the school.

A "Stepford" mother wearing a fluffy summer dress and pearls was walking with her son while her elbow was extended for her son to hold on to. The boy was at least 12. The whole vignette seemed to be something from the 50's except for the stay at home dad. "Pleasantville" I muttered, "just like the movie". With all children gone, the old man stood there looking at both long lines of cars that had accumulated. He smiled and began his slow walk back to the curb, still doing his best to hang onto the stop sign.

I drove on to find the center of town and tried to guess which street could be the one leading to Tarrytown. I eventually chose one to the left, and started another twisting drive through the hills and trees, never knowing truly if I would find my destination. I was eventually forced to pick another parkway heading south, towards what I thought would be Tarrytown, but I wound up in White Plains more frustrated than ever.

I returned north on the parkway, tried to calculate miles based on the map and took a different exit about the right distance. I took another road with no sign to the right, but again the road twisted and turned and I found himself back in Pleasantville for the second time. "Just like the movie" I thought, "You can't get out of this town."

Eventually, I found my way back to the reservoir north of White Plains and I knew Tarrytown was on the west side of the water, and that was all I needed to find my bearings. It was 9:00 AM. Just another short 2 hour commute to the office. "So much for starting early today".

The day went on uneventfully, although I picked up more work than I could complete which on the whole was pretty normal. There was simply an expectation that I like everyone else on the project team, work whatever hours were necessary to get the job done. The problem of course that everyone was working 70 to 100 hours a week including the multi state travel.

Towards the end of the day, I asked a local for directions to the airport from the office. Surprisingly, it involved only three turns, but I would have to keep a sharp eye out for the road signs as the street changed names frequently. It was only a 20 minute drive but with lots of twisting and turning. I returned my rental with the narrow seats and low ceiling and vowed I would never rent one again. "Obviously a car designed for women and small men". I continued the internal monologue and the criticism while I walked towards the terminal.

My flight was at 6:30, which would have placed him inside the great state of Texas sometime after 8:00 PM Central time. I looked for a place to sit and plug in my laptop since my battery was almost gone for the day, but there were no outlets to be had. I did find one of the few open seats near the women's bathroom, and I received sometimes strange, sometimes curious looks as the other gender strolled past me on their way to the "lounge" at the White Plains airport.

The minutes passed and the flight was posted as delayed, befitting the events and my luck for the week thus far. 2 hours went by as did the occasional updates telling the every growing and angry crowd virtually nothing except for the nasty weather that had shut down most of the East to West air traffic. My connecting flight was supposed to take off from Chicago O'Hare in about 2 hours, there was still a slim chance I would make my connector to Ft. Worth. A few minutes later, all hope was lost with an announcement that 11 PM would be the first chance the flight could take off.

I frantically tried to find a hotel room in Chicago somewhere, and everybody seemed to sense the same trouble because all cell phones were in use in the terminal. I worked my magic Gold number with Marriott, but there was nothing at the airport, nothing Downtown, nothing in Westchester, and nothing in Oakbrook near the company offices. The only thing available was a Fairfield Inn in Willowbrook, some 30 miles outside of the airport. "Yes, I'll take it... what was the address again?"

The flight did eventually leave around midnight, six hours after scheduled departure, and I could see the lightning flash through the clouds as they flew West at 33,000 feet. The storms were everywhere and the flight was bumpy. No sleep for anyone tonight. They all just held on for the roller coaster ride. 90 minutes later, hot sleepy and sweaty passengers made their way through a mostly closed and vacant O'Hare airport past terminal "L" and towards the exit.

I was feeling a little sorry for myself then. A personal pity party as it were as I contemplated the events of the week so far, but changed my mood as I rounded a construction barrier that partially blocked a darkened hallway. I first thought it was odd that the lights were off here, because there were about 200 people walking through the area, but then I and everyone else noticed why. Around the corner, in the darkened hallway started a sea of folding cots with stranded passengers filling both sides of the long hallway. "It had to be over 400 people" I thought to myself as most of the people were still awake and throwing a curious glance our way as we casually walked through their makeshift bedroom.

I held my head down and tried to provide these good folk what privacy I could. Almost at the end of the long rows of cots, I spotted a scared little girl about the same age as my youngest daughter. She had pulled up the small blanket up around her neck and watched us all look at her as we passed by. She wasn't near anyone who looked like her parents, and her cot was close to a security guard station. I wondered if she was a juvenile passenger, traveling alone and stranded alone at O'Hare. I could do nothing else but think of my kids and told myself I would never allow them to travel alone.

Walking even more slowly after the scene in the hall, I found the cab stand and asked to be taken to Willowbrook and the Fairfield Inn. Passing the cab stand the car stopped and the well meaning but non-English speaking driver said "widow bruck?".

It was an interesting ride, and we only had to stop for directions twice. The driver did manage to form the words telling me that it would be "time and a half" on the meter since they were leaving the city of Chicago. "Fine" I answered. "Whatever, just get me there". The meter ran up to $30, then $35 and I checked my wallet. There was exactly 64 dollars in cash. "Uh… how much farther do you think… I might need to find a bank or a cash station". "Almost" is all I got in reply.

One more stop at a gas station when the motel didn't pop up at the exit like we both had hoped. It turned out that it was just around the corner and the meter read $40. I paid the $60 after time and a half, and offered 2 dollars tip, leaving me 2 bucks for dinner. I checked in, wiped the sweat from my forehead, and bought a grape juice from the vending machine in the lobby.

It was 3:00 AM Thursday. I couldn't sleep. The shower washed the stickiness off my body, and there was hope the hot water would help me relax when I finally laid down on the medium density and not very high quality foam mattress. I set the clock radio alarm and then threw the comforter off the bed only to notice the blood stains that had seeped through the printed top, to the white underside of the flowered bed dressing. "Flowered top" I chuckled, "De-flowered top was probably more like it." "You've got to be kidding" I commented out loud and I kicked the comforter in the corner of the room. I checked the pillows carefully and the sheets as well, and tried to drift off to sleep finally for once this week.

The alarm went off at 6:00 AM and I just laid there listening to the shrill beeping pound in my head. I reached my right arm across the bed and starting pounding the top of the clock radio in a feeble attempt to stop the noise. Rolling over to focus my eyes on the nuisance I finally shut the alarm off. I gathered my thoughts and called American Airlines to find out what happened to my flight to Ft. Worth and how I could get there this morning. The flight was cancelled after all in Chicago, so I would have been stranded anyway. The next available flight was scheduled for 12:30 PM that day and I happily took the chance.

There was no sleeping now. I pulled out my laptop while I listened to the sound of the traffic coming from just past my front window, I hated these motels. The curtains had to be absolutely closed tight or anyone strolling past your front door could see you in your underwear. The laptop booted and I tried to catch up on the backlog of work that I would never finish.

It was a daily effort to figure out who to disappoint and who to complete work for. "It is what it is" I thought, typically waiting for my boss or a cast of others reset my priorities for me on a weekly and sometimes daily basis. The keyboard clicked quietly and the overhead light reflected heavily off keys "E R T S D F C and I O J K L N M". The heaviest used keys on my keyboard, my fingers had worn the textured surface of the keys down to a glossy shine.

The front desk called for a different cab this time, where I was guaranteed a rate of $24 and "Yes" they would accept a credit card. "Good… since I don't have any cash left" I explained to the dispatcher. It was 10:00 and I decided to iron my clothes before getting them all rumpled on my flight. It was a challenge working the ironing board since the iron was chained to the back of the board in an attempt to prevent theft, but it all worked out somehow. The cab ride was uneventful and I had plenty of time to get my boarding pass and find the gate.

The cots were all gone now, and I thought of the little girl I had seen just a few hours before. I wondered how my own children were doing, and felt guilty again for not being able to speak to them or my wife much that week. I remembered former bosses of mine making statements like "You knew what you were in store for when you joined this company…" , "This work involves travel… period". "Yeah", I thought to myself, "but I took two other management positions with this company in order to reduce the travel… but here we are again". "What city is this?".

An uneventful flight led to an standard but late landing at Ft. Worth and I made a bee-line towards the Ft. Worth customer offices. It was almost 4 PM, but I tried to find my contacts anyway and get as much accomplished as I could. I met the new program manager and got involved in a conference call that pushed the day past 7 PM. I found some of my company's technical team still hanging around their desks and asked if anyone had plans for the evening.

"No.. but we were thinking about catching another Rough Riders game tonight at 7:30".

"Lets do that!" I answered back.

"I need a diversion after this week... lets go for sure even if we're late."

Late it was. There are always difficulties getting out of a "needy" customer site. There is always another question. There is always another meeting. There is always another status update for the senior management team. There is always something that cuts into your personal time.

Four of the gang including me, headed to the hotel to drop off one of their troop and the remaining three headed off to the nearby mall that doubled as parking for the Frisco Rough Riders Double A Texas-League baseball team. "Park at the mall, take the bus to the game... I like this". The day had caught up with me, perhaps it was the whole week and while standing in the parking lot waiting for the shuttle bus, I doubled over in pain. "Guys" "I'm sorry and I hate to do this, but I have a problem and I need to get to a bathroom right now". The mall entrance was relatively close so we made our way to find facilities inside the mall.

I think I did everything but die in that rest room. My head was spinning, I felt nauseous, I was sweating everywhere and my stomach felt like someone was pounding it with a sledge hammer. It took 15 minutes for me to get rid of whatever sickness invaded me. I was still dizzy a little as I found my co-workers in the mall and they had already come up with an alternative plan for the evening.

"There's a Dave and Busters one level up in the mall, lets go there and catch the rest of the Piston's game".

"Shit, that was tonight?" I asked, losing all track of the day of the week.

"Yeah man, aren't you a die hard Detroit fan?"

"Well yeah of course" I replied. " I didn't know game 4 was tonight".

Dave and Busters was pretty busy, but the lower level, the non smoking level still had tables available around the center bar with the game playing on 4 tv's.

"Cool, up by 12 in the second" I thought. As the Lakers mounted a small comeback, the remaining people around the bar were whooping it up on behalf of L.A. while I was probably the only person in the bar actually from Detroit. Ben Wallace had a huge rebounding game and the rest of the new "bad boys" had huge games blowing L.A. out by 20 points.

Detroit went up 2-1 in the series and I thought "nobody thought we could do it… look at us now." It was much like the work that I and my team did week in and week out. "No respect until you show up and do the job".

I thanked the gang for sticking with me despite the change of plans and they were gracious in return, maybe because I picked up the dinner tab but it didn't matter. Despite how the evening started, I still needed the break from the week and it was good to detach from everything even if only for a couple of hours.

Back at the hotel around 10:00, I plugged in the laptop to take care of the pile of E-mail that accumulated throughout the week. Each one I opened seemed to increase the work load if not through a direct demand for my time, then through needing to re-direct the customer on poor assumptions or decisions they were making during the week. My stomach started to grumble again. "Shut up" I murmured at my own mid section. "There's nothing left in there for you to churn up and blow out".

It was 1:45 Friday morning. I still couldn't sleep. I had put the laptop away around midnight, and gave up on the news and HBO around 1:00, but the recent pattern of lying on my back staring at the dark ceiling had emerged again. I had tossed around for the next hour or so, pulling the non-fitted sheet corners out from under the mattress and making a general mess of the bedding. Somewhere after 3:00, I dozed off but it never got more serious that that.

6 AM and the start of the last day of the week. I felt again as if I had gotten no sleep at all. The day thankfully was going to be brought to a close early today as my flight was scheduled for 2:30 back to Detroit. It was an exercise in meeting schedules today, and once again nothing much was decided. "Meetings for the sole purpose of having meetings" I thought. "No decisions, no take aways, no customer responsibility, nothing to indicate any forward progress at all." I had a quick lunch with some of my team and they all started to head for home, wherever home might be.

The slow pace of Friday took me down a relaxed path and for the first time in 5 days the tightness between my shoulder blades eased a little. I finished as much of my e-mail as I could, packed my briefcase and headed for the airport.

It was a struggle to stay awake even with the bright sun pounding its way through the tinted glass of my third rental car this week. I fought the impulse to close my eyes all the way to the DFW airport. The lines at the airport weren't that bad and the forecast for weather in Ft. Worth as well as Detroit was good. I remained hopeful. I checked at the counter for the chance of an exit row seat, which helped me with the leg room desperately needed in coach. Again, another good sign. "Exit window is available" Valerie said from behind the counter. "Thank you Valerie" I said as she looked a little puzzled but then realized her name tag was the give away.

I made it through security this time without any added search and I was starting to feel good for the first time this long week. I noticed on my boarding pass that the exit row people get to board first. "Group 1" the ticket said, which was good considering the gate was jammed with people whose travel plans were also interrupted this week. "I'll at least have a place for my carry on luggage". When the announcement for Group 1 was made, I was fifth in line and ready with my boarding pass. "Please step to the side sir" was all the gate agent had to say and I knew what was to come.

The Last Strip Search

Even more invasive than the other times this week, the search station was located right next to the gate door without a privacy screen. As my dirty underwear and t-shirts were being tossed in my duffel, the entire line of people looked on.

I stood in front of the line of passengers 3 feet away with feet spread and arms outstretched while the security agent reached inside my belt line. I tried not to look at them, but they all looked at me trying to figure out why I was targeted for search. It was a little strange that a 6 foot 5 inch white male, a documented frequent traveler and a US citizen was chosen for the search while several other men and women from somewhere in the Middle East boarded without incident. I chastised myself for even thinking that way, "profiling" was not something I believed in, but things were definitely different now after 9/11.

I boarded last, because the search took 20 minutes, but I thankfully found room for my crushable duffel bag. This flight would be on-time and I started to count the minutes until I would be home. I thought about what I would be doing this weekend, I thought about my awaiting "honey do" list, I thought about how my kids would think of me years later after missing so much time at home. I remembered once introducing myself to a neighbor I had first met last year, and the neighbor had no idea that I actually lived there. "Yeah, I figured that some people would have picked me as the weekend boyfriend" I joked, but it was probably what they thought.

The 737 leveled off at 33,000, I closed my eyes for a little while and tried to detach from the entire week. I thought of the family cabin I hadn't seen for a while and hoped I could get there soon. I thought about hanging the hammock again now that the weather was getting warmer. I thought about some trees that needed trimming near the lake. I thought of the wind blowing through the trees and the branches over head swaying slightly in union with the hammock.

The sound of the wind in the trees was in my ears now and I could distinguish the higher pitched tones of the wind through Birch leaves and the deeper tones of the Oaks. I fought the impulse to doze off knowing the flight was only two and a half hours long but exhaustion got the better of me. I tilted my head towards the window, relaxed my jaw and felt my folded arms slowly fall into my lap. Finally, sleep.

Wrapping up in Texas

During the last few weeks of our strategy and planning assignment in Texas we were busy wrapping up documentation for review, edit and customer sign-off. We were doing our best to deliver as many recommendations as we could given the limited information available in some areas of the architecture.

About the same time the customer had also hired an independent management consulting group to help guide them through the pitfalls of a global software implementation, I guess. Interesting I thought because I thought that was why we were there. I was at odds almost immediately with their chief architect over their declarations of what was "best practice" and what software product should be used to resolve a business need.

"Its clearly an upper right quadrant product and should be the only one considered for this customer" They would say.

Try to follow me on this one. From my perspective "customer best practice" can mean just about anything. Customer best practice based on what you read? Or customer best practice based on the few customers you might have visited? And the whole concept of a "visionary" new software product, please. Research it further. If a product is visionary, but the only customer references you find from the vendor's website are "Bob's Car Wash, and "Mel's Internet

Service", is that all the affirmation you need to implement it for a multi-billion dollar company?

I was personally disappointed when the customer's senior management team started taking advice from people who had never implemented our architecture or integrated to our product lines. I'm sorry, I realize I sound like a PeopleSoft "bigot", but it is what it is. PeopleSoft experience should have been given the weight in the decision making process and it wasn't. This statement applies to technology, architecture, software products, and project management techniques. I of course wish them luck, but after more than a year and many millions of dollars invested, the actual project has yet to get past the strategy and planning phase.

Chapter Eighteen – Circling Back

After Texas, I took over the development of a services product that had been popular when PeopleSoft 8 first came out. It is what the company liked to call a "packaged service". It was in fact a full set of documentation templates, discovery session forms, sample graphics, budget calculators, training media, everything you would need to learn about and ultimately delivery that service to a customer. I had access to a small but talented team of technical folk and we merged heads on our collective customer experiences.

We figured about 2 months of development time for the new package and everyone on the project could effectively work from home. I like those projects. We decided to use some of the materials we had developed in Fort Worth, as those templates seemed to work very well when collecting data and documenting decisions. The work continued through the months of August and September.

More potentially bad news for PeopleSoft was published on September 9th, when the Department of Justice failed to win its claim that the Oracle/PeopleSoft merger would violate anti-trust law. Justice Vaughan Walker ruled that Oracle's acquisition would not violate such a law. Although it didn't mean the deal was done by any means, the stock market reacted in anticipation that PeopleSoft could fall victim to the buy-out. The mood in the company hit another low despite attempts by senior management to define exactly what the ruling meant.

It was a challenge in the field as well, trying to educate customers as to exactly what the Justice was saying in his opinion and ruling. Always on the front line, customers wanted us to explain how we were still fighting off Oracle's hostile takeover.

September ended slowly as we wrapped up our early development on the packaged service, but there was always the Iceland trip to look forward to. Don and I continued our weekly calls to exchange ideas on what to do and where to go. It was something that kept us going despite the dismal weather we were having this year in late summer.

The Circle

Which of course brings us back to the beginning of the story. After the long flight from Boston, Don and I were both crashed out in our mini-hotel room when we heard the news about Craig Conway's exodus and Dave Duffield's return. It was a constant topic of conversation that weekend in-between long moments of introspection about family and career. The weekend in Reykjavik went by too fast for words, and Van Morrison was awesome. I got a chance to take pictures with my new-but used medium format camera in black and white. The weather in Iceland that weekend was constantly overcast and rainy, but I managed to shoot two rolls of film for later development. All in all it was a therapeutic trip.

We returned to work the following Monday and I tried to catch up on all the news reports and e-mails about Craig and Dave. Early that week Dave issued an internal memo talking about why the decision was made and what his commitment would be to the board, he company and the employees. "DAD" as he would refer to himself (The insiders knew DAD = David A. Duffield), was back and it appeared back to stay.

The Oracle camp was silent over the Duffield news, and the CEO change eliminated a speculated position that Conway was blocking the takeover deal as part of some personal agenda. With Conway gone, no personal agenda.

October continued as any other month in the shadow of Oracle, but it was also time for PeopleSoft to post financial numbers for its operations in Q3. Quite unexpected, but PeopleSoft posted record numbers for Q3, with a number of deals going to new customers and earnings per share higher than anticipated by Wall Street. Congratulatory memos were issued internally to all groups for continued hard work and some smiles were even seen in the offices around the country. Maybe the cloud was lifting.

On November 1st, Oracle issued another offer for stock at $24 per share and had removed some of the restrictions that were inherent with previous offers. Still, it was less than the $26 per share offer made in 2003, and our current stock price then hovered just over $23 per share. They called it their "best and final offer" and it was set to expire at midnight on November 19th. We wondered if this was a "save face" kind of offer knowing it was unlikely they would be able to acquire the 51% of the stock they need to force the take-over. The patient and dedicated hopefuls that remained here at PeopleSoft, certainly hoped so.

I waited until the ninth of October, but I committed an expense felony and used company paper to print a 10 day tear-off countdown calendar. It simply read:

"10 Days Until Oracle Goes Away"

"9 Days Until Oracle Goes Away"

"8 Days Until Oracle Goes Away"

OK, you get the idea.

Focusing On a New Assignment

The following week I managed to grab two technical guys I respected in the region for some work at a government facility in Atlanta. I had hoped that the boys still in the office would fulfill tear-off duty for me on the countdown calendar since I was out of the office. The facilities in Atlanta were sparse, as I would expect a government facility to be, but there were no analog phone lines and no secure way to get outside the network. We were closed off from the outside world at least until we got back to the hotel. That week from news reports we learned of the growing likelihood that Oracle was going to acquire more than 50% of the shares in its tender offer.

Friday was deadline day and I boarded my flight back to Detroit in the early evening and arrived back home around 8 PM. My wife asked me if I knew anything, and of course the answer was as it had been for 17 months. "We don't know anything yet". We had a late dinner with the kids and then we went off to bed around 11:00, hoping for some sleep.

The morning yielded some bad news but there always seemed to be more to the story when it came to bad news about the takeover. As I learn more about hostile takeovers one day at a time, I realize that I may never know the whole story or all the chess moves to a game like this one.

Oracle was able to acquire 61% of the outstanding shares, but as it was explained by our CEO these are "tendered shares", not actual stock acquired by Oracle. PeopleSoft would appear in court the following Wednesday to seek an affirmation about the legality of their "poison pill" strategy. This strategy would prevent Oracle from actually "purchasing" a controlling interest of shares in the company. Even if the court upholds the poison pill strategy, Oracle could still fight for controlling seats on the board in the April 2005 stockholder meeting.

Monday continued with a flurry of e-mail and instant messages all from co-workers wondering what it all meant and asking what they should do. Later in the day I noticed a quote placed on my desk by someone in the Detroit office, and it was positioned on a table near my tear-off countdown sheet. It read *"It is a simple matter to drag people along, whether it is a democracy a dictatorship or a parliament. Voice or no voice, the people can always be brought to the bidding of the leaders. This is easy. All you have to do is tell them you are being attacked and denounce the pacifists for lack of patriotism. It works the same in every country.*" This was from Hermann Goering at the Nuremberg trials.

I just had to absorb that for a while. Was this person saying that the PeopleSoft leadership was using the same tactic? Accuse Oracle of attacking the company then count on its employee base for its patriotism? I remember a conversation I had with another employee a couple of months before who said, "If the price is right, hell I'll sell". "It's all about the money and that's the way its supposed to be". "Capitalism baby".

Somebody placed this Nazi quote in my work area, directed (I think) at my corporate patriotism. Maybe I was the fool but then again, maybe it wasn't just the money for me. Maybe I was interested in more than cashing out or just keeping a job. Although I felt like I held no misconceptions about reality when it came to hostile takeovers, I also thought there was still a way to keep the dream alive. This was a good place to be, and a profitable place to be a stockholder. Why couldn't everyone see that.

The mood in the office for the rest of the week continued to be sullen, with brief moments of humor to break up the thickness of the stress. We all talked about the upcoming court hearing on Wednesday, where PeopleSoft would be able to enact its poison pill strategy..... Or not that is. There were precedents to the strategy which allowed other companies to try and protect themselves. It felt to us like the court would uphold our strategy too. We would not know until late in the day Wednesday or worse, Thanksgiving morning.

Wednesday November 24, 2004

The day before the long awaited holiday break was supposed to the start of the Delaware court case to uphold the PeopleSoft poison pill strategy and prevent Oracle's acquisition of its tendered shares. Decision unknown, additional court dates set, Happy Thanksgiving.

New hearing dates were set for Monday December 13th and once again we remained hopeful that we would finally be able to defend our own interests against a jolly red giant. We told ourselves that it was not only good for the employees, but good for the long term stockholder as well. We had the products, we had the technology, we had the pipeline for future sales.

I chose to refocus on my laptop screen and count down the minutes before I left for my holiday turkey and tryptophan dreams the following day. The weekend came and went without much fanfare and I had to return to Atlanta on Monday to finish off the government work we had started the week before.

Aside from tripping over the customer's network cable and doing a "header" into the wall leaving me with a nice bruise above my eye (a first for me), the week started and ended without fanfare. I got a chance to work again with two guys who I respected and a project manager from Minneapolis who was always great to work with. It wound up to be a good week. We said our good bye's to each other Friday afternoon to start our respective weekends.

The week of December 7th was used for wrap up of the government documentation and a few other loose ends from November. We knew the poison pill court case would begin the following Monday, but internal news letters and notes of encouragement were strangely missing that week. It had become part of the weekly sometimes daily messaging since the Oracle news hit us almost two years ago, but the internal airwaves out of Pleasanton were silent.

My daughters both had Christmas holiday band concerts that week, and I was glad that I was home to see both the 6th grade "Cadet" band and the 7th grade "Concert" band. Not too many squeaks or missed notes this year, which alone was a sign of a successful concert. I remembered again how many of these things I had missed in the last nine years, and I watched how other parents would say hello to my wife but have no idea who I was. I just smiled and nodded.

That weekend was also my son's 16th birthday. All he really wanted for a party was to have a few friends over to play cards. Thanks to popular cable broadcasts of Las Vegas poker tournaments, all that these kids seemed to want to do was play "Texas Hold 'Em". What started out as three friends to fill the table, quickly turned out to be 16 kids piled into the basement at three tables where the overflow traded turns at video games.

They were well behaved I guess, for teenagers that is. The party that started at 5:30 with pizza, broke up around 11:30 with empty pizza boxes still laying on the floor of the basement. The throng of unemployed teenagers all left quietly, some even said "Thank you". I remember thinking later as I lay in bed. "I have a 16 year old son". "Oh my God". "When does Social Security kick in?"

It was a productive weekend around the Wortham house. Out of respect for our son's birthday, we usually held off on putting up holiday decorations until after mid-December, and this was the weekend to put it all up. Normally this fit in with the schedules of the rest of the people in our subdivision, but this year was more "proactive" than others.

Each year the lights seem to come out earlier than the last. This year, certain people had outdoor decorations up and lit the week after Halloween. Everyone else in the neighborhood seemed to jump on this bandwagon the second week of November. Our house was the last in the block to put up its lights on December 12. I was waiting for a nasty letter from the homeowner's association about my lack of Christmas spirit, I swear.

December 13, 2004

After a good weekend and thankfully an absence of thoughts about Oracle or my future, I slept in a little longer than I normally do but made it to my car by 7:30. On the trip to the office I listened to my local sports station to hear more details about how the Lions blew another lead on Sunday against the Packers. Damn that Farve guy, but you just have to love the way he comes back from being down.

Stuck in traffic, I got a cell call from my boss Matt who quickly asked me if I had heard the news yet this morning. "No, what news?". "As of this morning we've agreed to be taken over by Oracle." I remember sitting there silently for a moment, then re-focusing on the car in front of me because I needed to stop quickly. "Fuck" was the only thing out of my mouth. In retrospect, not a great show of class in front of my boss but it was about the only thing that could sum up my mood right at that moment.

Matthew went on to explain what he knew which wasn't much, but it was somewhat of a personal call so it was appreciated. The remainder of the drive left me numb, just numb. No other word aside from that four letter utterance upon first hearing the news could describe how I was feeling. The four letter word was my blue collar reaction, but the numbness was physiological. The tips of my fingers were tingling and void of any feeling.

There were only a few people in the office that early but first reactions maybe told the story best. I found one of my Director buddies down the hall and all she could do was provide a blank stare at me without saying a word. We both knew what shock looked like. I wondered if my mouth was still open. Early claims from Oracle's management team about immediate reductions in work force, something like 6000 employees came to mind again. We all just speculated, what else could we do.

A somber internal e-mail was sent by Dave Duffield just after 9AM Eastern, clearly after a sleepless night for many in California. It said everything we didn't want to hear but the most revealing and emotional line talked about how he was "saddened by this outcome". A corporate message was later sent indicating a 12 Noon Eastern all employee conference call.

At Noon, each officer took a chance to speak on the call. Dave was first but handed over the call to Kevin Parker for the financial details. Aneel took over and eventually had to hand the call over to Phil Wilmington. Ultimately everyone leading the call was overcome by emotion at some point in reading the prepared text.

Some of the people in our office conference room were in tears, while the rest of us stared at the table or glanced around the room to share a moment with somebody we knew. It was the second most emotional meeting I have ever attended, preceded only by the "9-11" anniversary ceremony in New York. After an announced moment of silence, our customer's PA-system read the names of their employees who had died in the crash the previous year.

There was now some hope of employee retention based on statements made by the Oracle team, but most of us knew that these were all unknowns. It would be a fast merger as both companies had committed to completing the transaction before the end of the year. This statement seemed to us to be a little bit of a pipe dream given the JD Edwards merger took us months to complete and those circumstances were friendly. Who really knew.

In some ways it was a relief to hear the news, whatever the news actually was. It had been 18 months of not knowing, and despite the fact that we still would not know our employment fate, at least we knew what entity won the battle.

In the end it appeared that it was all about the money. I recalled a line from the movie Wall Street, where the main character had declared something like "Greed works". It was a small number of people or commercial entities who wanted to cash out and take whatever profits they could get. $24 wasn't enough, but $26.50 apparently was. Oh by the way, Merry Christmas, Happy Chanukah, Peaceful Kwanzaa.

Closure

At some point, there needed to be a wrap to this story if for no other reason than the Oracle offer had hung over us day to day, week to week and month to month without end. I watched as our future was threatened and our customer base was filled with doubt about our longevity and their previous investment in our software. Our other main competitor, SAP in Germany, seemed to take advantage of the whole situation here in America.

While two American giants fought and struggled to keep their footing, SAP ate up the market share of "concerned customers" with both feet planted firmly on the ground. I remembered a quote made by a consulting friend from Morocco, "When two elephants fight, its only the grass that suffers". I love that quote. "Call me Kentucky Blue".

In case of any layoffs, I was confident enough in my technical and managerial background to be able to find work somewhere, even if it wasn't exactly what I wanted to do. For me, it was all about wanting to hold onto a good thing, with a good product and for a good company. There seemed to be so few of them left in the marketplace. Many employees were contemplating leaving PeopleSoft early, just to be able to close that chapter in their lives and unload the stress of "not knowing".

My family remained as concerned as I was. We had already cancelled a vacation we had been planning for the last two years, and other purchases we needed to make were on hold for the time being. Cheryl had told the kids enough for them to understand some of the ramifications of my predicament, but of course they would never know it all. We simply did what most Americans did when faced with a potential financial problem. We did our best to plan.

Many of us at work had talked about how much they were sick of the whole IT industry and the direction every company seemed to take when it came to knowledgeable labor. "Cheaper must be better". The permanent positions available with most companies were reducing year over year, and the consulting trade was becoming just as tough to carve out your own specialty niche. I actually thought about opening a Hallmark franchise. I thought it would be cool to have Cheryl and the kids work there and I could build displays for holiday knick-knacks. Some of the other gang at work talked about owning bars or restaurants for their next move but we all knew we would be back doing what we had always done.

With PeopleSoft, I was grateful for many things I had learned here and many things I was rewarded with during my tenure. It had been the most dynamic 9 years of my life affecting pretty much every part of my life. The laundry list of learning experiences rolled on the more I thought about my tenure at PeopleSoft. I started to summarize one-liners as if I were to appear on stage with a microphone.

Random Ramblings on Things That I have Learned

The relationship with my wife and kids was strained because of the time away, but somehow we learned to make our family stronger.

I built relationships with some of the most intelligent and interesting people I've ever met, calling many of them "close friends" to this day.

Travel opened up new experiences for me and I drew a new respect for other cultures both here in America and abroad.

If a customer asks you to take them to a bondage club for fun, reconsider the concept of delivering to this new level of customer service.

Based on disagreements and confrontations with many customers and some co-workers, I had learned that there are always three sides to any story: What the first person thinks, what the second person believes and finally the actual truth.

Never eat salsa from a squeeze bottle with a narrow tip, or sit across from someone who does.

A good leader beats a good manager any day, at least when it comes to getting things done. Its amazing to me how few people will stand up to take the lead in any situation, especially when the situation desperately calls for leadership.

Single source vendors, "prime" contract holders, and master service agreements always seem to cost the customer more money in the long run.

Good technology people seem never to be appreciated for their insight or experience, but are absolutely relied upon to solve the crisis they worked so hard to avoid in the first place.

The typical three hour European "family" dinner is a social and epicurean joy that American people need to experience only once to fall in love with. Rib-lets and fries in 30 minutes at the local bistro don't count as "dinner".

Never argue with a German holding an engineering degree. He will be right until you prove him wrong. (Apologies here, just my experience)

Fast food on the road… forget about it. The "burrito supreme" is an oxymoron, and the "regular taco" is more appropriately named. Find a restaurant with at least a real knife and fork.

Find a hotel chain with free high speed internet access and room service. While on the road, you will need both when your work day ends at 10:45 PM.

If you hear gunshots outside of a restaurant with a fenced, guarded parking lot, it may not matter how good the steaks actually are inside. Perhaps it shouldn't be the cholesterol you worry about.

Buying technology for technology's sake is an easy pit to fall into based on poor consulting advice. Wise customers will choose technology in order to solve business problems, not to provide functionality that no one asked for.

Minor league baseball and hockey games were the some of the best entertainment for the buck. And the hot dogs and beer are pretty good too.

Customer service and the concept of maintaining a personal relationship is an art that most people do not do well, and also what customers seem to need most.

Finally, always bet the maximum on the slots in Vegas. "YOU ALWAYS BET THE MAXIMUM". (Picture older woman in house coat next to a overflowing ashtray here.)

Epilogue

As for my family, despite all the travel and time away from home we seem to have maintained a pretty good relationship. All time at home for me is important time with my wife and kids. Many of my previous local friendships have suffered because of this, but since home time is limited to a local office week here and there or weekends only, there isn't much time for friends or sometimes extended family. So instead of "bowling night" or "boys night out", it's a Disney movie with the girls, or a hockey game with my son or even a dinner out alone with Cheryl.

I always found it interesting that as complex as my work life had become, it forced me to simplify my personal life. By that I mean that when the work is done and I am back at home, my time seems dedicated to my wife and children. No other commitment or social engagement seems to get in the way. And when I am away, I look forward to getting home and taking them out to dinner to hear about the week. I look forward to going up north to the family cabin where there isn't cable TV, or wireless connectivity, or high speed internet anything. Where we play board games and shop for fishing lures and just spend time with each other.

I realize that a claim of a "functional family" it is a very subjective statement to make, but we do see other families in restaurants, at the mall, and in our neighborhood. We hear the gossip (mostly because our neighbor seems to know all and tell everything), and for some reason our little family enterprise seems to be working pretty well. I truly hope so. We work hard at trying to make it all work some how. Yes from time to time I get to play with my friends, and sometimes out of the country, but it is an infrequent thing and Cheryl always gets the travel pay-back. "Shopping in Chicago again this year?" I feel my husband training kicking in. "Yes Dear".

Maybe it works too because I still like what I do for a living. Consulting on the road and managing people is difficult but it is rewarding as well. There is a social aspect to this kind of professional life and considering that we are placed in weekly situations where the city and the people are "new", it is never boring. Given all that weekly instability, the rock that I've always needed to lean on is that family. My family.

New Employment?

I cannot say how long I will be with our new employer, maybe that remains in the hands of the new board and the new management. I still want to help our customer base and do what I do, "wacka do, wacka do". Maybe this will be a turning point in my career where I simply move on to something radically different. Handmade canoes, or black and white photography, or hmm, what was that other thing? Oh yeah, writing.

I'm not sure I could work for any other consulting company, based on what I've seen elsewhere and the way certain firms treat their employees. I expect that I will always be able to do something that adds value to this company and for our customers. I expect that I'll continue to like the people here. I expect that my wife will continue to support me with my career despite the fact it sometimes takes me away from home. "Right Dear?" "er I mean, Yes Dear".

Consulting was and is a hard road to travel. Travel being the key word here. It is a profession that is painfully challenging but rewarding at the same time. Not that I would recommend it for the 40+ year old with kids, it seems to be a career for the young and the endlessly energetic. Having said that, it also seems to be the over 40 crowd with experience and the ability to manage the stress of a complex situation. If you're young and willing to travel, I might recommend it as a first job to gain that experience. Over 40? It has to be the right thing for you and your family.

And happiness? It seems to be right there under your nose. Not that I buy into all sorts of eastern philosophy but I like one concept of being happy with what you have rather than unhappy because of what you don't have. The concept of needs versus wants. Actually that seems to fly in the face of capitalism a bit doesn't it? When it comes to your profession and your family, life seems to become less stressful when you focus on the good fortune you already have. Hopefully the good health of your family, the good people you work with, your life experiences are all good things. I will concentrate on the good things and continue to give thanks for what we have, especially through the holidays of 2004 and the 2005 New Year. I encourage you all to do the same.

Best of luck to you, in all your personal and professional journeys.

Peter Wortham

About the Author

Peter Wortham is a pseudonym for an author living in South Eastern Michigan and working for PeopleSoft the past 9 years. Coming from a banking and insurance background, Peter was leading development teams and major IT projects at the age of 24. Moving to PeopleSoft in 1995, Peter quickly improved his consulting skills and was promoted into the consulting management ranks. Peter is married and lives with his wife and three children in Michigan, while still traveling extensively for the PeopleSoft/Oracle management team. Peter brings this experience and a humorous slant on consulting and family life to the pages of this book.